INTRODUCTION

Offshore oil and gas are the most important British natural resources to be discovered this century. They provide energy and essential chemicals for our transport, industry and homes, and earn valuable tax and export revenues to support the British economy. How have we come to be so fortunate?

Oil and gas are formed from the remains of ancient plants and animals. That they exist as workable deposits is the result of several vitally important factors coming together at the right time and place:

1　An environment in which plants and animals flourished, and which ensured the burial and preservation of their body tissues after death.

2　Burial to sufficient depths to convert their remains to oil or gas.

3　The right sort of rocks to store and seal in the oil and gas – porous and permeable rocks make good reservoirs, but sealing layers must be impermeable.

4　The right structures in these rocks to trap the oil and gas. Since oil and gas are less dense than rocks and water they tend to migrate upwards, eventually escaping at the surface.

5　Trap structures must have formed before the oil and gas are generated or have entirely migrated away.

The discovery and development of our offshore oil and gas is a triumph of painstaking detective work and bold construction initiatives undertaken in a difficult and dangerous environment. Their great economic importance provided the incentive to overcome the difficulties and dangers and has inspired enormous advances in the science and technology of offshore exploration and production. The operating companies have taken huge financial risks to meet the great costs of offshore development, and thousands of workers daily endure the perils of the North Sea to produce our essential oil and gas.

1　Brae 'A' flaring during a production test

ORIGINS OF OIL & GAS

Oil and gas are derived almost entirely from decayed plants and bacteria. Energy from the sun, which fuelled the plant growth, has been recycled into useful energy in the form of hydrocarbon compounds – hydrogen and carbon atoms linked together.

Of all the diverse life that has ever existed comparatively little has become, or will become, oil and gas. Plant remains must first be trapped and preserved in sediments, then be buried deeply and slowly 'cooked' to yield oil or gas. Rocks containing sufficient organic substances to generate oil and gas in this way are known as *source rocks*.

Dead plants usually decay rapidly and are dispersed, but in areas such as swamps, flooded forests and sheltered lake- and sea-beds, vast amounts of plant material accumulate. Bacteria breaking down this material may use up all the available oxygen, producing a stagnant environment which is unfit for larger grazing and scavenging animals. The plants, bacteria and the chemicals derived from their decay become buried in silts and muds and are preserved (fig 2).

Whether oil or gas is formed depends partly on the starting materials (fig 2). Almost all oil forms from the buried remains of minute sea-algae (fig 3) and bacteria, but gas forms if these remains are more deeply buried. The stems and leaves of buried land plants are altered to coal, but little oil is formed; with deeper burial they produce gas. Britain's offshore oil and gas originate from two sources. Gas from beneath the southern North Sea and the Irish Sea formed from coals which were derived from the lush, tropical rain forests that grew in the Carboniferous Period, about 300 million years ago. Oil and most gas under the central and northern North Sea formed from the remains of planktonic algae and bacteria that flourished in tropical seas in the Jurassic Period, about 140 million years ago. They accumulated in muds which are now the prolific Kimmeridge Clay source rock.

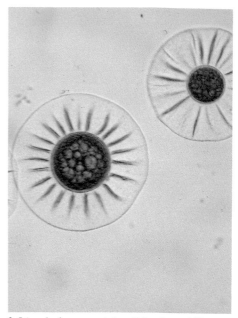

3 Live plankton containing oil droplets ×150

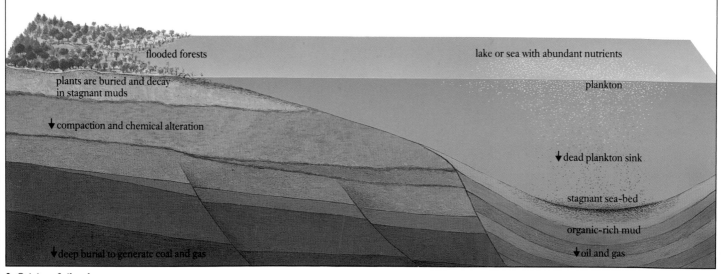

flooded forests

lake or sea with abundant nutrients

plants are buried and decay in stagnant muds

plankton

↓compaction and chemical alteration

↓dead plankton sink

stagnant sea-bed

organic-rich mud

↓deep burial to generate coal and gas

↓oil and gas

2 Origins of oil and gas

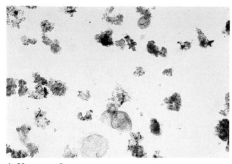

4 Kerogen fragments

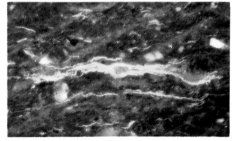

5 Fluorescing oil in a clay source rock

The processes of oil and gas formation resemble those of a kitchen where the rocks are slowly cooked.

Temperatures within the Earth's crust increase with depth so that sediments, and any plant material they contain, warm up as they become buried under thick piles of younger sediments (fig 6). Increasing heat and pressure first cause the fats, waxes and oils from algae, bacteria, spores and cuticle (leaf 'skin') to link and form dark specks of *kerogen* (fig 4). The cellulose and woody parts of land plants are converted to coal and woody kerogen.

As the source rock becomes hotter, long chains of hydrogen and carbon atoms break from the kerogen, forming waxy and viscous *heavy oil*. At higher temperatures, shorter hydrocarbon chains break away to give *light oil* and then gas. Most North Sea oil is the more valuable light oil. Coal and woody kerogen yield mainly methane gas, whose molecules contain only one carbon atom. Gas from the southern North Sea is methane.

When the source rock starts to generate oil or gas it is said to be *mature*. The most important hydrocarbons are gas, oil, oil containing dissolved gas, and gas condensate. Gas condensate is a light oil which is gaseous at high underground temperatures and pressures; it is the most important product in some North Sea fields.

In the North Sea (fig 6) oil forms at 3–4.5 km depth, gas at 4–6 km. At greater depths, any remaining kerogen has become carbonized and no longer yields hydrocarbons. Burial to these depths occurs in areas where the Earth's crust is sagging (p 7). In the central and northern North Sea the oil source rock is buried in a deep rift valley (pp 16–17). In the southern North Sea, coal-bearing rocks formed the floor of a basin which filled with younger sediments (fig 28, p 13). These processes continue today: the North Sea floor is still sagging so new areas of source rock are reaching depths where they too are generating oil and gas.

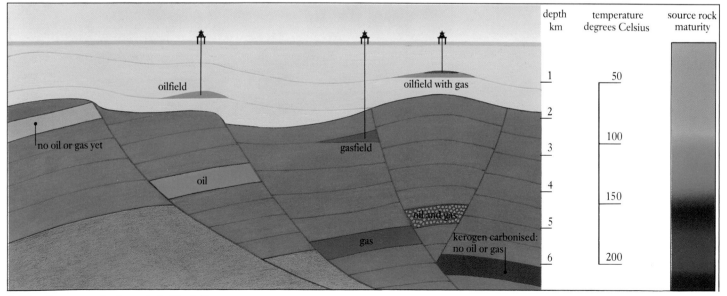

6 Oil and gas generation beneath the North Sea

MIGRATION

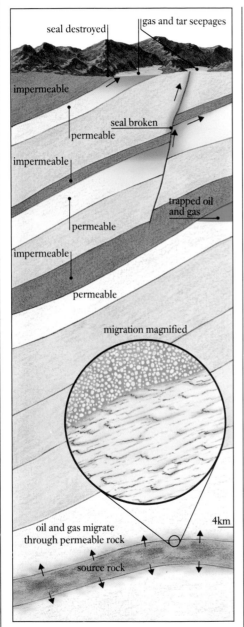

7 Migration of oil and gas

Labels on figure 7:
seal destroyed
gas and tar seepages
impermeable
seal broken
permeable
impermeable
permeable
trapped oil and gas
impermeable
permeable
migration magnified
oil and gas migrate through permeable rock
4km
source rock

Much oil and gas moves away or *migrates* from the source rock. Migration is triggered both by natural compaction of the source rock and by the processes of oil and gas formation. Most sediments accumulate as a mixture of mineral particles and water. As they harden to become rock, some water is squeezed out and dispersed; if the rocks contain oil and gas, this is also expelled. As hydrocarbon chains separate from the kerogen during oil and gas generation, they take up more space and create higher pressure in the source rock. The oil and gas ooze through minute pores and cracks in the source rock and thence into rocks where the pressure is lower (fig 7).

Oil, gas and water migrate through *permeable* rocks in which the cracks and pore spaces between the rock particles are interconnected and are large enough to permit fluid movement (fig 8). Fluids cannot flow through rocks where these spaces are very small or are blocked by mineral growth;

8 Open pore spaces in permeable sandstone ×120

9 Pore spaces filled by mineral growth ×90

such rocks are *impermeable* (fig 9). Oil and gas also migrate along large fractures and faults which may extend for great distances.

Oil and gas are less dense than rock and water so tend to migrate upwards. Much oil is dispersed as isolated blobs through large volumes of rock, but when larger amounts become trapped in porous rocks, gas and oil displace water and settle out in layers due to their low density (fig 10). Water is always present below and within the oil and gas layers, but has been omitted from most of our diagrams for clarity.

Migration is a slow process, with oil and gas travelling perhaps only a few kilometres over millions of years. But in the course of many millions of years huge amounts have risen to sea floors and land surfaces around the world, escaping as gas and tar seepages (fig 7). Liquid oil seepages are comparatively rare; most oil becomes viscous and tarry near the surface as a result of oxidation and bacterial action.

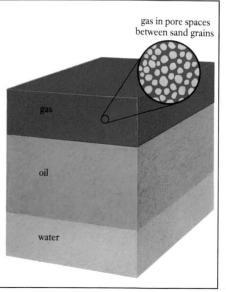

10 Gas, oil and water in sandstone

Labels on figure 10:
gas in pore spaces between sand grains
gas
oil
water

Oilfields and gasfields are areas where hydrocarbons have become trapped in permeable *reservoir rocks*, such as porous sandstone or fractured limestone. Migration towards the surface is stopped or slowed down by impermeable rocks such as clays, cemented sandstones or salt which act as *seals*. Oil and gas accumulate only where the seal and reservoir rocks are in the right shapes and relative positions to form *traps* (fig 12), in which the seal is known as the *cap rock*. The migrating hydrocarbons fill the highest part of the reservoir, any excess oil and gas escaping at the *spill point* (fig 11). Gas may bubble out of the oil and form a gas cap above it; at greater depths and pressures gas remains dissolved in the oil (fig 11). Since few seals are perfect, oil and gas escape from most traps. In many fields incoming oil and gas balance this loss, as in the Brent and Ekofisk fields in the North Sea. Gas migrates and escapes from traps more readily than oil, but the salt layers beneath the southern

North Sea have proved a very efficient seal because salt contains no pore spaces, and fractures reseal themselves.

Figure 12 shows the main types of trap. *Structural traps* are formed by Earth movements which fold rocks into suitable shapes (a) or juxtapose reservoir and sealing rocks along faults (b). Traps may also form when rocks are domed over rising salt masses (c). *Stratigraphic traps* originate where a suitable combination of rock types is deposited in a particular environment (d), for example, where a reservoir rock of permeable river sand is sealed by clays which accumulated in the surrounding swamps. In reality most traps are formed by a more complex sequence of events, and cannot be classified so rigidly. For example, in (e) the reservoir rock was first folded and eroded, then sealed by an impermeable rock which was deposited later over the eroded structure. Where a particular set of circumstances has combined to produce a group of oil- or

gasfields with similar trap structures or reservoir rock, this is termed a *play*. There are several important plays in the North Sea, which are described on pages 14–19.

In order to trap migrating oil and gas, structures must exist before hydrocarbon generation ceases. In some parts of the North Sea trap structures existed 125 million years ago, but were not filled with oil until 100 million years later. The rocks beneath the North Sea are sinking only a few millimeters in ten years, so cooking and generation of hydrocarbons is very slow.

All hydrocarbon fields form by a chance combination of events that produces the right sorts of rocks and structures, together with the right timing. The forces that shaped Britain's offshore hydrocarbon fields also created oceans and transported Britain from south of the equator to its present latitude. The origin of these forces and their effect on the crust, both globally and around Britain, are described in the next section of this book.

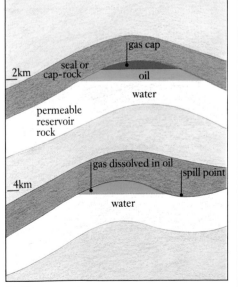

11 Traps

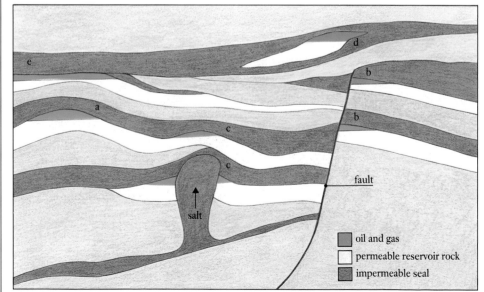

12 Trap structures

FORCES THAT SHAPE THE EARTH'S CRUST

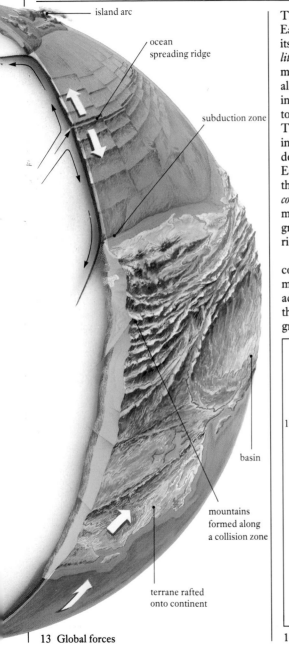

island arc

ocean spreading ridge

subduction zone

mountains formed along a collision zone

basin

terrane rafted onto continent

13 Global forces

The semi-molten white-hot interior of the Earth is in constant motion. This transmits itself to the more rigid outer layer, the *lithosphere*, which is also constantly on the move (fig 13). New lithosphere is created along mid-ocean ridges where molten rock is injected, cooling to form new *ocean crust*, the top layer of this young lithosphere (fig 14). The lithosphere moves away from the ridges in the process of *sea-floor spreading*, and is destroyed wherever it slides back into the Earth, along *subduction zones*. Since it is thicker and lighter than the oceanic crust, *continental crust* is not subducted and so is mostly much older than oceanic crust. The great slabs of lithosphere between mid-ocean ridges and subduction zones are called *plates*.

The complex interactions of oceanic and continental lithosphere powered by plate movements are called *plate tectonics*. In addition to the opening out of ocean basins, the main effects of plate tectonics are the growth and break-up of continents.

Continents grow by formation of new continental crust and by the addition of *terranes*, which are pieces of continental material and ocean island arcs formed elsewhere and rafted into older continents by sea-floor spreading. These collisions telescope the continental crust and produce mountain ranges. Conversely, where the spreading process locates itself under a continent, the continent may eventually split apart. A new ocean will form between the rifted parts which may then travel long distances as passengers on the moving plates. The rate of growth and horizontal movement of plates is anything from about 2 cm to 10 cm per year, about the same as one's finger-nails. The drifting of rifted continents may carry them through several climatic zones, for example, from equatorial humid through tropical arid to temperate and arctic, over tens or hundreds of million years (fig 15). This is of great importance to the generation and trapping of oil and gas, as are

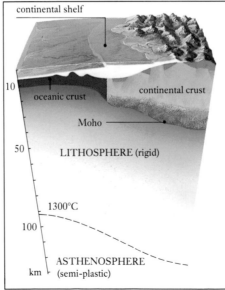

continental shelf

10

oceanic crust

continental crust

Moho

50

LITHOSPHERE (rigid)

1300°C

100

ASTHENOSPHERE (semi-plastic)

km

14 Section through the lithosphere

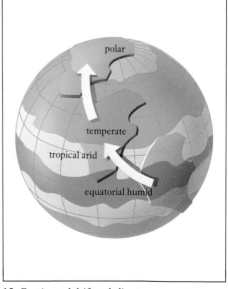

polar

temperate

tropical arid

equatorial humid

15 Continental drift and climate

the structural disruptions brought about by plate tectonics.

Large areas of the continental crust are covered by layers of sedimentary rock which are thickest in the middle of *basins*. Nearly all oil and gas is found in such basins, which are formed over many millions of years by stretching of the crust combined with sagging. The North Sea is a classic example. Most basins have a two-tier structure: the lower tier is block-faulted whilst the upper tier is a simple sag (fig 16b). There are different theories to explain basin formation. The lithosphere may stretch uniformly like toffee, fracturing the upper brittle layers into tilted blocks, then sag as the underlying, partly-molten layer (asthenosphere) cools down. Alternatively the entire lithosphere may be detached along a huge low-angle fault (fig 16a) to which curved (listric) block-faults are linked. The reality may be a combination of stretching at depth with detachment high up in the lithosphere.

Compression of the upper continental crust by plate tectonic mechanisms results in buckling and telescoping of rock layers to form fold and thrust belts (fig 17). The telescoping is often related to a deep detachment, above which a stack of thrust sheets piles up. Large masses of lightweight granite give buoyancy to the crust. The *highs* that result may be marked by reduced deposition of sediments or actual emergence and erosion. Beyond the thrust belt, rock strata may undergo compression. This tends to expel the contents of basins upwards and outwards in a process termed *inversion*. The expulsion often takes place along the same listric faults that guided the basin's development. Basin inversion is a very important mechanism in gas and oil fields (p 23). It may create good structural traps for oil and gas, and may prevent 'over-cooking' of the source rock. However, it may also permit the escape of hydrocarbons or cause erosion of source rocks or reservoir rocks.

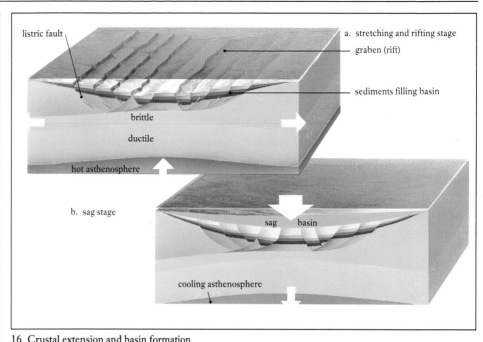

16 Crustal extension and basin formation

17 Crustal compression and basin inversion

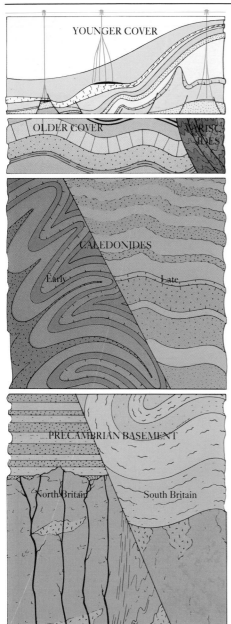

18 The main crustal units of Britain

Although the plates of the Earth's crust are in constant motion, the strongest effects of plate tectonics occur at peak times with longish periods of relative tranquillity in between. Rock strata accumulate in these tranquil periods and are disrupted and deformed by later plate collisions. This causes thickening of the continental crust which is thereby elevated as mountain ranges. The mountains are eroded down and deposition of sedimentary strata resumes. The crust is thus built up of units, some highly deformed, others hardly disturbed. These units are stacked above or against each other. The junctions of the units are either old erosion surfaces or large dislocations.

There are five major units building the crust under British seas (fig 18). At the top is the *Younger Cover*, ranging in age from Permian to Recent (fig 19). It fills the basins containing our oil and gas.

Below is the *Older Cover* comprising Devonian and/or Carboniferous strata between 400 and 300 million years old. In southern Britain, rocks of this age are strongly deformed and cut into slices which have been stacked above each other to form the Variscan fold-belt or *Variscides*. Under much of the North Sea, the Older Cover forms an 'underlay' resting on early Palaeozoic strata folded 410 million years ago (the late *Caledonides*) and mainly late Precambrian strata folded and metamorphosed 510 million years ago and earlier (the early Caledonides).

Precambrian rocks older than 600 million years form the foundations of the whole structure. In the north are ancient crystalline rocks with a flat cover of less ancient Precambrian strata involved in adjacent early Caledonian folding. In the south, the Precambrian basement was not formed until the very end of the Precambrian Era.

The map (fig 20) shows the outcrops of the main units of the continental crust in

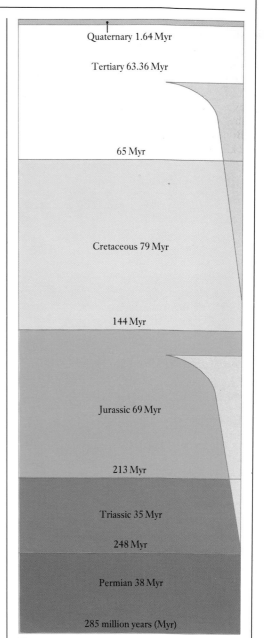

19 Younger Cover time scale

North-West Europe. It is immediately obvious that the British Isles are surrounded by seas mainly underlain by the Younger Cover formations. The map shows the location of the main depressions in which the Younger Cover was deposited. Some of these exceed 10 km in depth. Over 'highs' such as the Mid-North Sea High and the London–Brabant Platform, the thickness diminishes to 2 km or less.

The Older Cover is extensively exposed on the sea floor only in the western Irish Sea and in the Orkney-Shetland region. Subsea outcrops of the Variscides border the Western Approaches and Celtic Sea basins. The Caledonides have a very restricted subsea outcrop consistent with their tendency to form upland areas of Britain and Ireland. The ancient Precambrian basement of North Britain – the 'Lewisian Gneiss' – is widely exposed whereas the young Precambrian of south Britain has almost as few subsea exposures as on land.

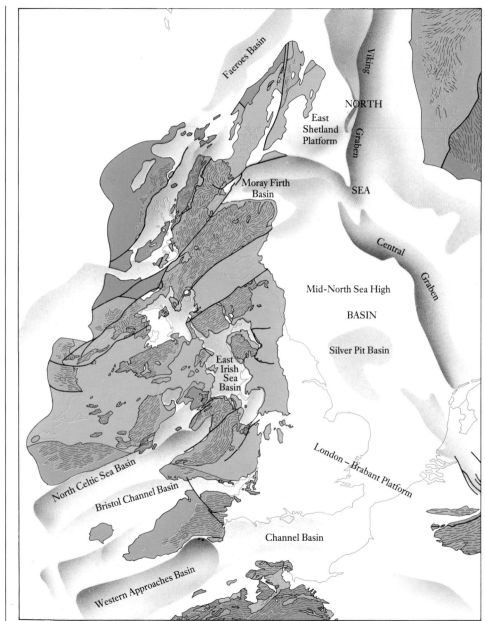

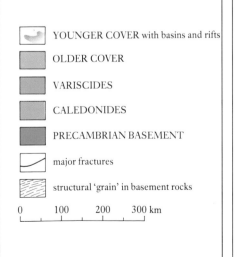

YOUNGER COVER with basins and rifts

OLDER COVER

VARISCIDES

CALEDONIDES

PRECAMBRIAN BASEMENT

major fractures

structural 'grain' in basement rocks

0   100   200   300 km

20 The structure of Britain and the adjacent continental shelf with, above, a section from the Western Approaches to the Hebrides

FOUNDATIONS

The Precambrian basement forms two-thirds of the British crust. The northern half (fig 21) was part of the supercontinent of *Laurasia* (Canada, Greenland and Scandinavia) and has had a long, eventful history dating back nearly 3000 million years. Most of it has been repeatedly reformed at high temperatures and pressures. The southern half, composed of less altered rocks, was a northern 'outpost' of the supercontinent of *Gondwana* (Africa, South America, India and Antarctica). The two halves were brought together when the *Iapetus* Ocean closed, creating the next tier, the Caledonides. The early Caledonides formed before Iapetus closed. They are metamorphic rocks with an incredibly complicated structure caused by repeated compression. The late Caledonides are sedimentary and volcanic rocks folded after Iapetus closed. The Norwegian Caledonides are intermediate in age and have a thrust-sheet structure unlike the British Caledonides. The North Sea is underlain at depth by all three types of Caledonides.

The Caledonian plate collisions in the Devonian Period produced a land of lakes and wide river plains in which the Old Red Sandstone was deposited. A seaway extended across southern Britain and into central Europe at this time. In the Carboniferous Period a warm coral sea flooded the Old Red Continent, but as the continent drifted northwards towards the equator, the sea retreated and humid swamplands developed, clothed in luxuriant vegetation. These, the coal forests, were periodically flooded when the polar ice-caps melted. Late in the Carboniferous Period, the southern seaway closed and Laurasia became fused to the main mass of Gondwana, with the formation of the Variscan mountain ranges – the Variscides – along the collision zone (fig 24). By the Permian Period all continental masses had come together to form the vast supercontinent of *Pangaea* which continued to drift northwards.

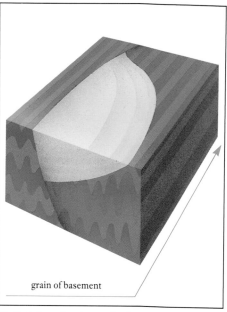

grain of basement

23 Relation of basin trend to basement 'grain'

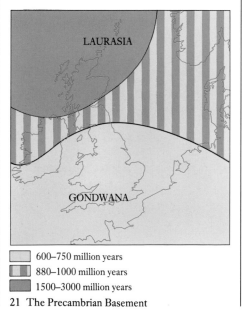

- 600–750 million years
- 880–1000 million years
- 1500–3000 million years

21 The Precambrian Basement

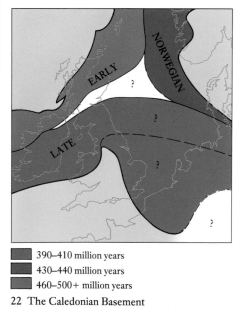

- 390–410 million years
- 430–440 million years
- 460–500+ million years

22 The Caledonian Basement

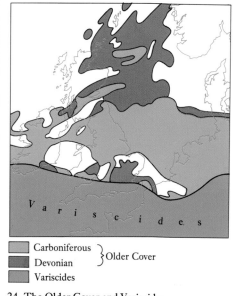

- Carboniferous } Older Cover
- Devonian }
- Variscides

24 The Older Cover and Variscides

The forces that created Britain's offshore oil and gas basins and their structures were related to global plate tectonics (pp 6–7). Widespread basin formation occurred in the Permian period during the crustal subsidence that followed the Variscan folding. Often it simply added to pre-existing Devonian and Carboniferous basins of the Older Cover, and carried on into the Mesozoic and Tertiary Eras. Many of these superimposed basins seem to be aligned along the dominant structural 'grain' in their basements (fig 23). In the North Sea, the two main Permian basins running east-west are most likely related to the adjacent Variscan fold-belt (fig 25).

Whether or not the Permian basin subsidence represents the earliest sign of the opening of the Atlantic Ocean is highly controversial, but by the middle of the Jurassic Period the opening was well under way. It advanced northwards in the Cretaceous Period and in the early Tertiary Period the North Atlantic rapidly opened out. During a crucial Jurassic and Cretaceous phase of widespread crustal tension, the crust around Britain rifted. Located over rising columns of hot mantle, these rifts ('failed arms') never became oceans but were the initial phase in the two-stage mechanism of basin formation (p 7). Tearing movements also produced local 'pull-apart' basins. Compressions originating in the plate collisions that produced the Alps caused uplift and erosion (inversion) of basin contents.

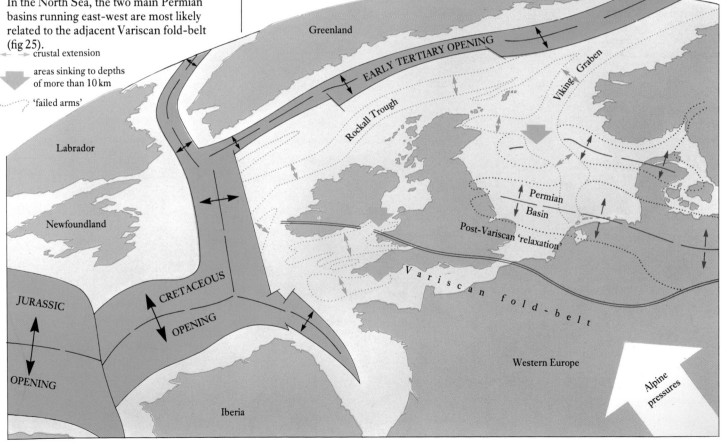

25 Basin-forming forces acting on the British crust 285 million to 60 million years ago

WESTERN BASINS

A number of deep basins are located off western Britain from the Shetlands to the Western Approaches. Most began life in the Permian Period and the deeper ones continued to subside throughout the Mesozoic and Tertiary Eras.

'Half-grabens' formed by the 'trap-door' mechanism are common; large basins like the East Irish Sea Basin (fig 26) are compounded of several such half-grabens. Apart from Morecambe Bay, Britain's second largest gasfield, the hydrocarbon prospects have been somewhat discouraging. The Jurassic, where present, comprises shallow-water and alluvial sediments not ideal as source rocks. The Morecambe Bay gas comes from underlying Carboniferous coals and has been trapped in Triassic sandstones beneath impermeable mudstone and salt. The flow of gas is hampered by growth of secondary clay mineral between the grains of the reservoir rock below a certain depth. Drilling on the slant is needed to tap the extensive but shallow clay-free reservoir.

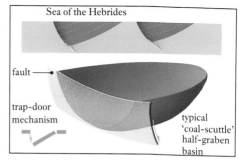

27 Half-graben basins

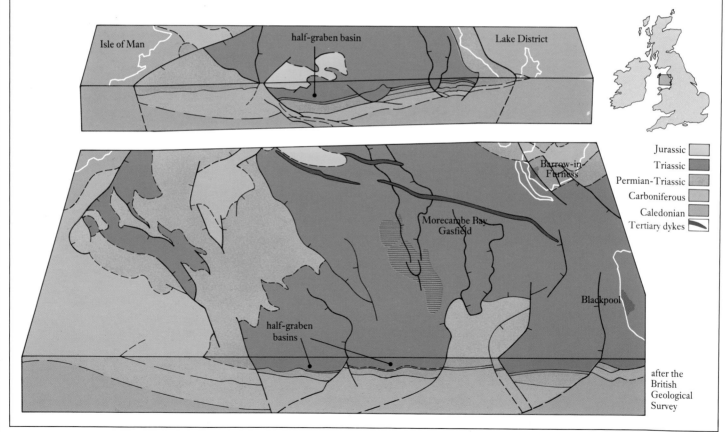

26 The East Irish Sea Basin

Concealed beneath the blanket-like sag of the North Sea Basin is a complex of older basins and rift valleys (grabens) between elevated 'highs' and platforms. In the southern and central North Sea, the thickest sediments are in the two Permian basins and in the deep Central Graben. The northern North Sea is dominated by the sediment-filled Viking Graben. Were the graben empty, Mount Everest would just about fit into it – upside-down. Fig 28 shows the Permian basins and the buried grabens as they would appear if all rocks younger than 285 million years were stripped away. The early rifting stage of basin formation lasted until about 125 million years ago, and was followed by the main sagging stage. The Earth's crust is thinned symmetrically across the entire width of the northern North Sea (p 7). Two further important influences on the structure in the central and southern North Sea are deformation produced by mobile salt masses (fig 29), and 'inversion' of basins accompanied by erosion (p 7, fig 17).

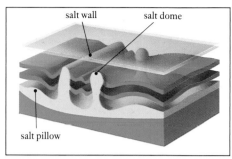

29 Effects of salt movement

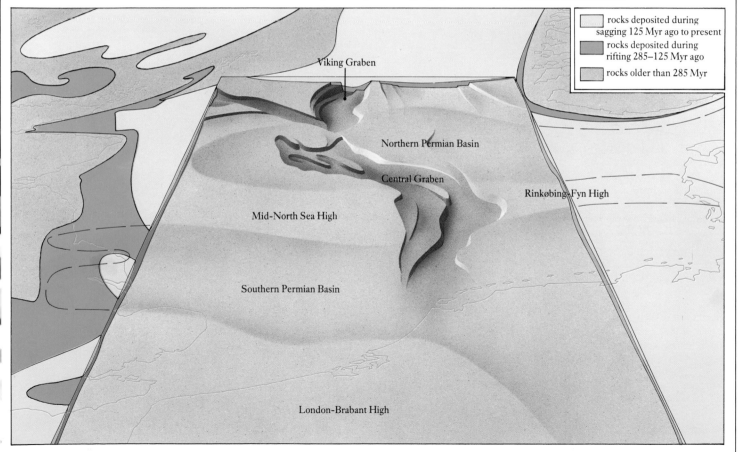

	rocks deposited during sagging 125 Myr ago to present
	rocks deposited during rifting 285–125 Myr ago
	rocks older than 285 Myr

Viking Graben

Northern Permian Basin

Central Graben

Rinkøbing-Fyn High

Mid-North Sea High

Southern Permian Basin

London-Brabant High

28 Anatomy of the North Sea

THE NORTH SEA – THE SOUTHERN GASFIELDS

Almost all of the hydrocarbon fields in the southern part of the North Sea are gasfields. From 285 million years ago, this basin area was set amid the vast Pangaea continent. During its slow drift northwards across the equator (p 6, fig 15), the environment in the basin gradually changed as the climate altered. Seas and lakes came and went, surrounding hills were worn down and the basin continued to deepen. During this continual change, a series of rock layers and structures have followed each other in a way which, by chance, allowed the creation and storage of natural gas beneath the area presently occupied by the southern North Sea (fig. 32).

Fig 30a shows the environment 300 million years ago when the area lay over the equator. Lush, swampy rain forest covered the flat lands across the area of Britain and the North Sea. These were the Carboniferous coal forests, the remains of which now provide Britain's coal and natural gas resources. Layers of vegetation were periodically submerged as the land sank and sea level fluctuated. As they became more deeply buried the plant layers were converted to coal seams in beds of shale and sandstone; these rocks are the Coal Measures, which underlie the area shown in fig 31a. Below about four kilometres, the Earth's heat drives gas from the Coal Measures coal and shale. Continued sinking, particularly along the area of the North Sea, has caused the generation of large quantities of gas; parts of the Coal Measures were at the right depth for gas generation to take place over 140 million years ago. The process is continuing still in some places. Because gas permeates upwards, it would all be lost at the land surface or sea-bed but for the existence, in certain areas, of overlying rocks which contain and seal in gas, efficiently trapping it. These rocks are the products of varying desert environments which affected the mid-

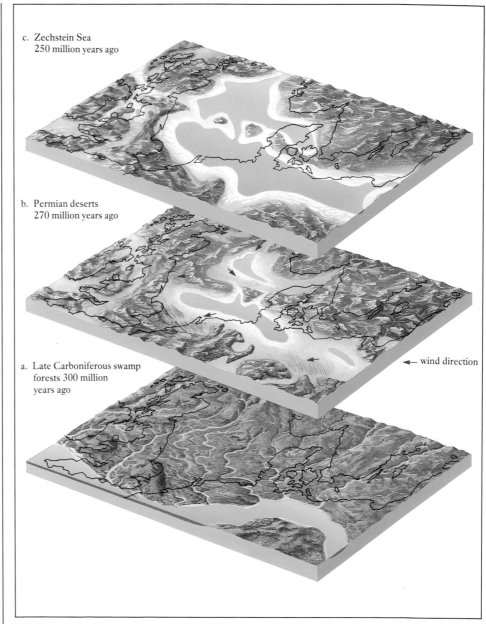

c. Zechstein Sea
250 million years ago

b. Permian deserts
270 million years ago

a. Late Carboniferous swamp forests 300 million years ago

← wind direction

30 Environments that produced the rocks of the southern gasfields

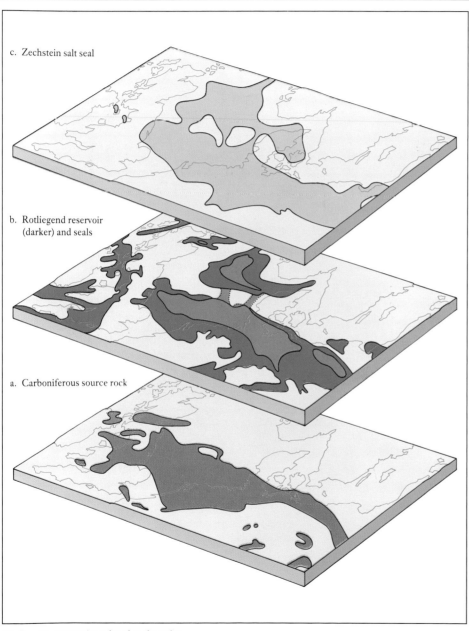

c. Zechstein salt seal

b. Rotliegend reservoir (darker) and seals

a. Carboniferous source rock

31 Source, reservoir and seal rocks today

continental regions of Pangaea (p 10) during the Permian and Triassic Periods. The trap structures formed later, many around 70 million years ago.

By 270 million years ago, a desert lake in the south was bordered by massive sand dunes cut by wadis (fig 30b). These sands built up to a 300 metre-thick sandstone formation while the area subsided. Some of the dune sands form the most permeable parts of the Rotliegend Sandstone Group (fig 31b) and hold much of our natural gas. 250 million years ago, the inland Zechstein Sea deposited salt layers over the dune sands (fig 30 and 31, c). These layers are now thick beds of salt which, in places, act as a gas seal.

Fig 31 shows how part of the southern area contains Permian Zechstein salt above Rotliegend Sandstone which, in turn, lies over Coal Measures. Wherever these three layers lie one above the other there is a chance that gas may be held, so long as the rocks are in the form of a trap (see pages 5 and 23).

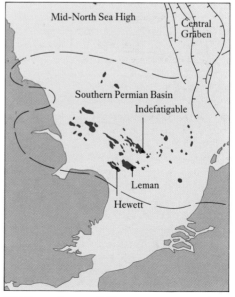

Mid-North Sea High
Central Graben
Southern Permian Basin
Indefatigable
Leman
Hewett

32 The southern gasfields

OIL & GAS FROM THE BURIED RIFT VALLEY

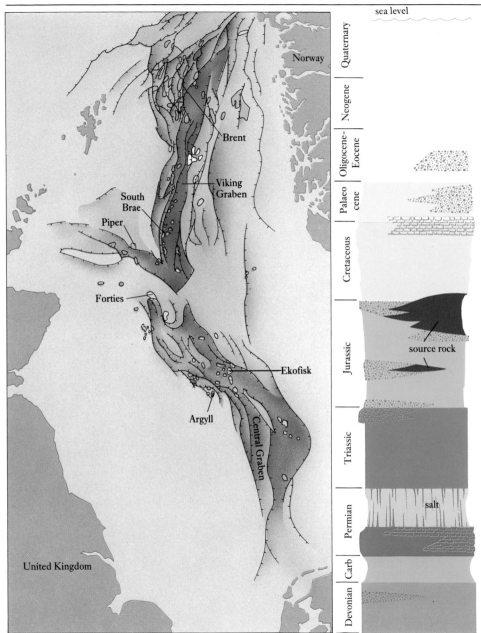

33 The rift valley, with the main oil- and gasfields coloured according to reservoir rock age

The story of oil and gas in the northern and central areas of the North Sea is dominated by the geological history of the buried rift valley, or graben. The rift is seen in the map on the left, which shows the present-day shape of the surface of all rock that is more than 285 million years old. During this last 285 million years, since the start of Permian times, subsidence along the line of the rift valley created a changing pattern of land, lake and sea environments, and influenced the thickness and type of sediments that accumulated, the depth to which they are now buried, and the trap structures that formed. Consequently, hydrocarbon deposits have been trapped in a much greater variety of rocks and structures than in the southern North Sea. Much of the oil and gas is found in sandstones that are less than 200 million years old. The main sandstone and limestone reservoir rocks are shown in green ornament in the column on the left.

Rifting movements affected this area for more than a hundred million years. They were most intense during the Jurassic Period, and Jurassic rocks provide the most important oil source and reservoirs beneath the North Sea. The main source of oil and gas in the area is the 140 million year-old Kimmeridge Clay (above, right). The most prolific oil-bearing reservoirs beneath the northern North Sea are the Jurassic 'Brent Delta' sands (fig 34 and p 33, fig 73). Brent Delta sediments also contain coal seams derived from vegetation on the swamps. These are the source of some of the gas now trapped in the area. The sediments that built up the delta were transported northwards by rivers draining volcanic uplands which had risen up at the junction of three 'arms' of the rift valley (fig 34). As large as the Nile Delta, the Brent Delta is now buried and broken into a series of tilted blocks (fig 36) which act as traps where overlying rocks seal down oil and gas (p 20, fig 42).

The Kimmeridge Clay is particularly rich in hydrocarbons along the line of the rift valley. This is because the slow subsidence of the rift helped to set up the right environment for a rapid build-up of thick mud layers, rich in planktonic algal remains, on the deepest parts of the sea-bed (fig 35). Climate and sea conditions were ideal for the massive growth of 'blooms' of plankton. Dead plankton sank in vast numbers, and the sea-bed bacteria feeding on their remains made the mud stagnant (p 2, fig 2), so that particles from the plankton cells were preserved in it and slowly buried. The buried mud became compressed to form the shale known as the Kimmeridge Clay. The thickest mud layers were deposited over the rift and have since subsided deep within the rift, heating up slowly as they became more deeply buried. The Kimmeridge Clay has been scalding hot for millions of years, generating first oil and then gas (p 3, fig 6). Fig 37 shows the areas where it is mature and generating oil and gas right now.

deep sea, thick mud
land
swamp

35 The Kimmeridge Sea 140 million years ago

low
degree of maturity
high

37 Kimmeridge Clay maturity today

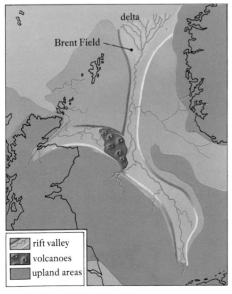

delta

Brent Field

rift valley
volcanoes
upland areas

34 The Brent Delta 170 million years ago

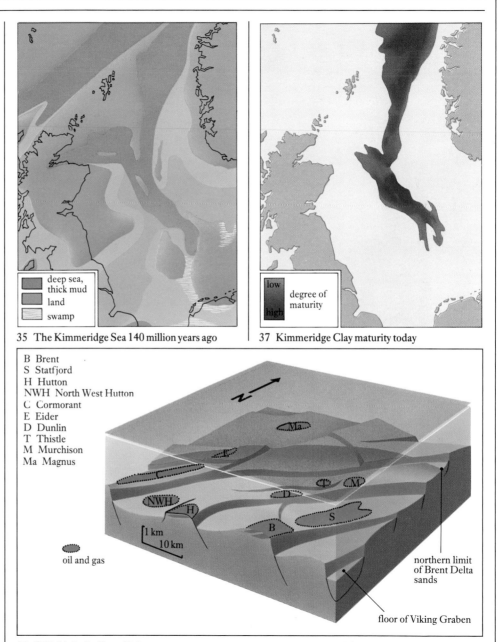

B Brent
S Statfjord
H Hutton
NWH North West Hutton
C Cormorant
E Eider
D Dunlin
T Thistle
M Murchison
Ma Magnus

oil and gas

1 km
10 km

northern limit of Brent Delta sands

floor of Viking Graben

36 Oilfields in the Brent Delta area

OIL & GAS FROM THE BURIED RIFT VALLEY

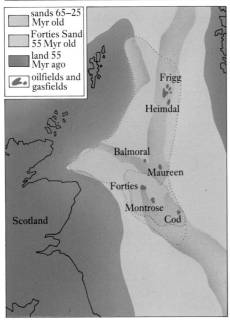

Legend:
- sands 65–25 Myr old
- Forties Sand 55 Myr old
- land 55 Myr ago
- oilfields and gasfields

Frigg
Heimdal
Balmoral
Maureen
Forties
Montrose
Cod
Scotland

38 The Forties Field area 55 million years ago

Most of the sandstone reservoirs in the northern North Sea were originally parts of river deltas (p 17) and 'submarine fans' – lobes and sheets of sediment which were re-deposited from slumping and flowing masses of unstable sea-bed (fig 39). Many important oil and gas occurrences are in Jurassic rocks of this type (p 21) but some of the largest are in submarine fan sandstones which were deposited much more recently (fig 38 and p 21, fig 45).

Around 150 million years ago, in Jurassic and on into early Cretaceous times, parts of the sea floor repeatedly sank. The great rift valley system, or graben, was rapidly subsiding. Beneath the sea, the Earth's crust continued to fracture along huge faults, and large blocks dropped down and tilted to form long ridges along the sea floor.

These movements continually triggered the slumping of soft sediments into the deeper troughs. Unstable areas of sea-bed would start to shift until rock fragments and particles were carried away across the sea floor as fast-flowing currents of watery sediment. These settled out as fans or more widespread sheet-like deposits. Coarse, sandy rubble was dropped near steeply-sloping sea floors. Channels and fans of sand and silt spread out further across the sea floor, building thick layers out from the submarine ridges (p 21). Some of the sandy rocks which were laid down in this way are permeable enough for oil or gas to flow through them with ease. These rocks now hold oil and gas pools in trap structures, such as those at Brae, Galley, Claymore and Magnus fields. The traps formed well before the oil and gas migrated in from the Kimmeridge Clay above or around them.

The North Atlantic Ocean was opening rapidly around 50 million years ago and through this time of great crustal activity the area from the Scottish Highlands to the Shetland Islands was uplifted (fig 38), causing rivers to erode and move huge amounts of sediment. Unstable masses of sand and silt built out across the surrounding shelves, while reactivation of older faults continually triggered great flows of sediment from the edges of the submarine shelves, out across the deeper sea floor underlain by the rift valley (fig 38). Submarine channel deposits and fans (inset) built up into widespread layers of sandy sediment. These have hardened to form beds of silt and sandstone with shale layers. Where they have become parts of suitable trap structures, as at Forties, Montrose, Frigg and Cod fields, they may hold considerable quantities of oil and gas which have migrated upwards from the deeply-buried source rock. Under some parts of the North Sea area, they have then migrated almost horizontally some tens of kilometres along sandy layers until they either escaped or became trapped.

Natural storage of oil and gas beneath some parts of the North Sea depends upon

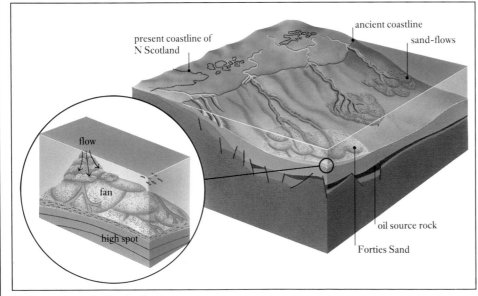

present coastline of N Scotland
ancient coastline
sand-flows
flow
fan
high spot
oil source rock
Forties Sand

39 Deposition of the Forties Sand

the presence of thick layers of salt, especially those laid down in the tropical sea during the Permian Period, around 250 million years ago. In the arid climate, rapid evaporation of the continually inflowing seawater resulted in the build-up of more than 2000 metres of salt.

In the central North Sea area, trap structures have been created where low-density salt layers have risen through overlying rock (fig 40 and p 13, fig 29). Some of these structures have trapped oil and gas (p 22), particularly within the Central Graben, where the chalk in the Norwegian and Danish sectors has been fractured and domed by rising salt.

In southern areas of the North Sea, however, salt layers of the same age act as hydrocarbon seals. Here, the source rock and the main reservoir rock lie beneath the salt (p 23, fig 48) and are not affected by its movement. Fractures in the salt heal by salt-flow, so the rock makes an excellent seal.

Chalk acts as an oil seal in some areas and as a reservoir rock in others. Normally, its permeability is low – oil will not flow through it. Chalk mainly consists of tiny mineral crystals formed by algae which drifted as plankton in the seas more than 60 million years ago. The crystals, made of calcite, collected on the sea-bed as white, limy mud which hardened to form chalk rock. Where deeply buried, its minute pore spaces become naturally cemented and the rock hardens. Deep within the Central Graben, however, some chalk is much more permeable than normal and contains oil and gas. A crucial factor was that of sediment-slumping on the sea-bed. Movements across the Central Graben rift-edge caused the sediment to flow and re-deposit as a very porous, watery slurry. In places, the pores were filled with oil at a high pressure before the crystals could become cemented into a tight mass (p 22).

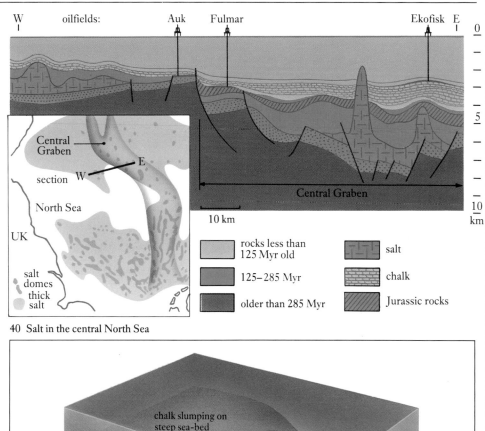

40 Salt in the central North Sea

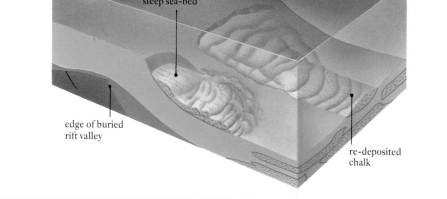

41 Chalk deposition in the Central Graben

19

A CLOSER LOOK AT SOME NORTH SEA FIELDS

Brent Field, discovered in the far north of the area in 1971, contains oil and gas within tilted layers of sandy rock. 170 million years ago, these layers were part of a river delta (pp 17 & 33). Since then, the tilting movements, associated with the rifting Viking Graben (pp 13 & 16) have been followed by a long period of sagging. Muddy sediments – including Kimmeridge Clay, the source of the oil – have draped across the tilted blocks (p 17, fig 36), filling the subsiding troughs between them, and sealing the eroded upper edges of the sandstone layers (fig 42) to form traps. Much later, oil was expelled downwards into the sandstones from the thick mudrock, now deeply buried within the troughs. Oil has migrated up the tilted sandstone layers to collect in the crests. Some of the gas came from coal within the delta sediments. Oil is still migrating through the area. The sandstone layers, each more than 200 metres thick, have held over 500 billion litres of oil, for millions of years, within an area of 17 by 5 km.

 Piper Field, discovered in 1973, lies at the edge of an arm of the buried rift valley (p 16). Oil in this field is trapped within a tilted sandstone layer cut by faults. The sandstone was deposited 145 million years ago, during late Jurassic times, as sand bars around a series of river deltas. Kimmeridge Clay source rock overlies the oil-filled sandstone, acting as part of the seal. However, the trapped oil is mostly derived from within the rift valley, on the south side of the field, where the source rock is thicker and hotter. The oil has migrated to the field area at some time after a mudrock seal was laid down 70 million years after the sand. Oil is prevented from leaking out at the faults, or from the eroded edges of the sandstone, by this seal. About one cubic kilometre of the sand is filled with 150 billion litres of oil over an area of 30 square kilometres.

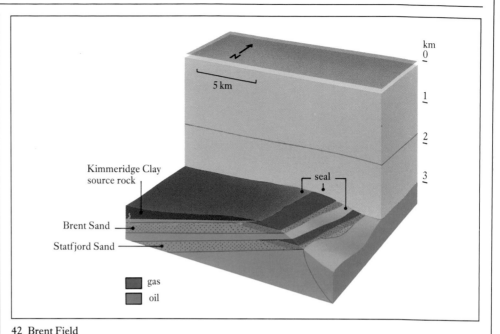

42 Brent Field

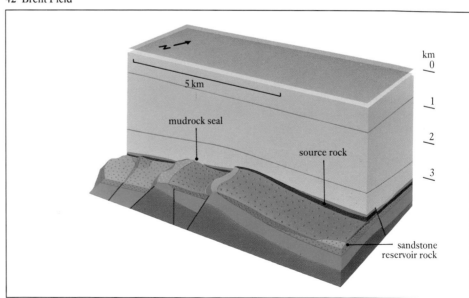

43 Piper Field

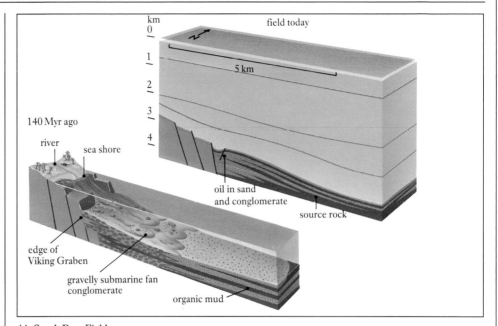

South Brae Field contains oil and gas in the sandy debris which accumulated at the foot of a steep submarine slope (p 18). This coarse sediment was deposited under the western edge of the buried rift valley, in the southern part of the Viking Graben. At this time, 140 million years ago, the organic mud of the Kimmeridge Clay source rock was being deposited across the area. During episodes of instability, fan deposits of rock fragments and sand spread out from the submerged, rifting edge of the graben, while organic mud deposition was confined to the floor of the undersea rift valley away from the steep edges. Thus the reservoir rocks are now found as sand sheets and wedges of conglomerate – pebbles and boulders in sand – interlayered with black, oily mudrock. Here, therefore, the reservoir rock is of the same age as the source rock. The oil and gas are still moving from their mudrock source into the adjacent reservoir rock. South Brae Field was discovered in 1977, its oil being found deeper down than in most North Sea fields, within a maximum thickness of over 500 metres of reservoir rock. The oil is hot, gassy and corrosive.

Forties Field, discovered in 1970, has held well over 500 billion litres of oil in its sandy reservoir rock. This rock was deposited as a submarine fan sediment 55 million years ago (figs 38 and 39, p 18). At Forties Field, these sandy layers have draped and sagged across a 'hump' in the underlying rocks. The Main Sand reservoir is composed mostly of sandy sediment deposited as a submarine fan, while much of the separate Charlie Sand accumulated within feeder channels of sand flowing across the sea floor. Oil has migrated upwards from the Kimmeridge Clay source rock within the buried rift valley, and has then travelled along the sand layers. Some of the migrating oil has then been trapped within the dome-shaped beds of sandstone above the underlying hump.

44 South Brae Field

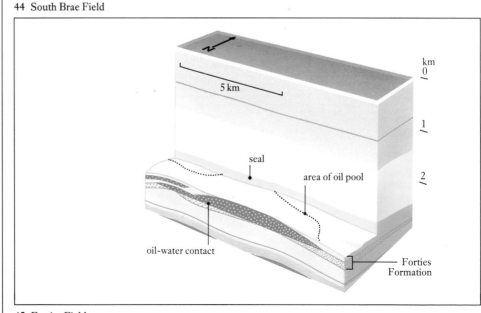

45 Forties Field

A CLOSER LOOK AT SOME NORTH SEA FIELDS

Argyll Field, discovered in 1971, lies on the western edge of the buried rift valley – the Central Graben – in the central area of the North Sea. Its oil lies within reservoir rock which is much older than the source rock. Migration along the rift fault zone brings oil from the depths of the graben; the reservoir lies within a spur-like block of rock which is fractured and gently folded. The block is also faulted on its western side, preventing oil leakage. The whole block is draped by a seal of much more recent shale and chalk, trapping the oil within an area of thirteen square kilometres. The reservoir rocks are mainly layers of desert sediments. The lower layers are Devonian rocks around 360 million years old. Above them, Permian dune sands with alluvium and lake-bed sediments are overlain by limestone, all around 250 million years old. Deposited as limy mud-flats around an inland salt sea, the limestone makes an unusual reservoir for this area. Some of its minerals have been dissolved out, leaving cavities which now hold oil.

Oil and gas were discovered at *Ekofisk Field*, in the Norwegian sector, in 1969. They are contained within chalk, a rock made of particles formed by sea life more than 60 million years ago. It is not normally a good reservoir rock. Much of the chalk in this area, however, is free of clay impurities and, moreover, was slumped and re-deposited across the deeper sea floor above the Central Graben area (p 19). This left chalk layers with a porous, open texture. The pores were later filled with high-pressure oil and gas, preventing minerals from subsequently blocking the pores. Thus the pores remain open and the rock stays permeable to the flow of hydrocarbons. Salt movements in the area fractured and domed the chalk (p 19), creating pathways for hydrocarbons to migrate up to the reservoir, as well as the structure to trap them. Gas has leaked from the trap into the shales overlying the reservoir.

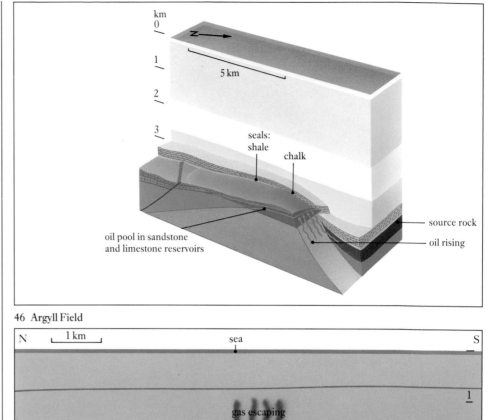

46 Argyll Field

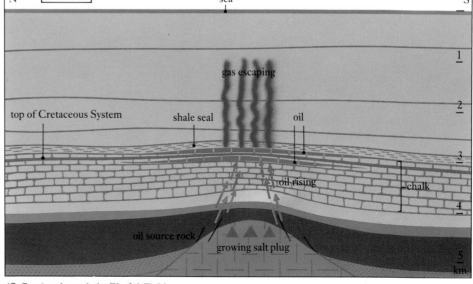

47 Section through the Ekofisk Field

Leman Field was discovered in 1966. It is the largest gasfield in the southern North Sea, underlying an area of more than 30 by 10 km off the Norfolk coast. The Leman Bank area is one of a number of places across this part of the North Sea beneath which the chance factors for the origin, migration, containment, sealing and trapping of gas have come together in the circumstances described on pages 14 and 15. The origin of the gas is in the coal-bearing shales of the 300 million year-old Coal Measures. The thick, Late Permian salt beds in this area form a very efficient seal above the excellent reservoir rock of dune sands laid down in Early Permian times, around 270 million years ago.

By 140 million years ago, certain areas of the Coal Measures source rock would have reached a temperature of around 130°C, at a depth of about 4 km, a sufficient level of heat to generate gas. Beneath Leman Field, however, the crust has since been pushed upwards, lifting the Coal Measures back out of the gas-generating zone. It is possible that the gas has either migrated from flank areas into the resulting dome-shaped trap (below) or it may already have migrated into the reservoir rock, dissolving in the water within the rock. It would then have bubbled out of solution when the pressure dropped during uplift. At nearby *Hewett Field* (below, left) these same upward movements allowed gas to escape past the Permian salt seal. It then migrated towards the surface but was trapped again in two sandstone reservoirs of Early Triassic age, around 240 million years old. These have seals of shale and a thin salt layer.

The main Permian reservoir sandstone was deposited in a great desert basin which covered the area between the eastern margin of England and the Russian-Polish border. Known as Leman Sandstone in the UK and Slochteren Sandstone in the Netherlands, this rock is part of the Rotliegend Sandstone Group (p 15). It is the most important gas reservoir rock in the southern North Sea.

However, as this sandstone becomes more deeply buried its reservoir capacity diminishes because the pores between the grains become blocked with mineral deposits. Uplift has not restored the gas-reservoir capacity of the rock.

In 1959, gas was discovered in Rotliegend Sandstone at Groningen in the Netherlands. The field later proved to be very large. With a thick salt seal above it, the sandstone was found to contain highly permeable dune sands which hold much of the gas. Such an important discovery led exploration geologists to follow the dune sands westwards along the direction of the prevailing wind which formed those dunes millions of years ago. This brought their attention into the area now covered by the southern North Sea – to a new region of the Earth's crust waiting to be searched, its stuctural pattern and geology discovered and its past environments deduced – the era of North Sea exploration had begun!

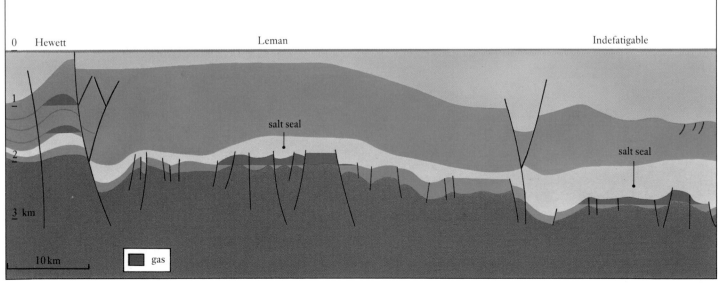

48 Section through the southern North Sea gasfields.

DISCOVERING THE UNDERGROUND STRUCTURE

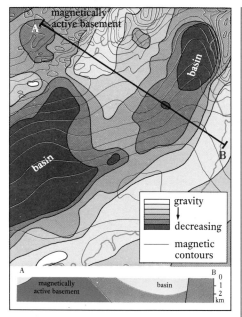

49 Gravity and magnetism of Cardigan Bay

Large-scale geological structures which might hold oil or gas reserves are hidden from sight by sea and by non-productive rocks. Geophysical methods can penetrate them to produce a picture of the pattern of the hidden rocks. Relatively inexpensive *gravity* and *geomagnetic* surveys can identify potentially oil-bearing sedimentary basins, but costly *seismic* surveys are essential to discover oil- and gas-bearing structures. Sedimentary rocks are generally of low density and poorly magnetic, and are often underlain by strongly magnetic, dense basement rocks. By measuring 'anomalies' or variations from the regional average, a three-dimensional picture can be calculated. In fig 49, magnetically active basement rocks at the surface show a busy magnetic signature in the top left, whilst the central area with quiet magnetics and low gravity values is a basin with up to 11 km of poorly magnetic sediments which might repay seismic survey.

Detailed information about the rock layers within such an area is obtained by deep echo-sounding, or *seismic reflection* surveys. In offshore areas these surveys are undertaken by a ship (fig 52) towing both a submerged *air or water gun array* to produce short bursts of sound energy, and a *streamer* several kilometres long containing up to 240 *hydrophone* groups. The hydrophones collect and pass to recorders echoes of sound from reflecting layers. The depths of the reflecting layers are calculated from the time taken for the sound to reach the hydrophones via the reflector; this is known as the *two-way travel time* (figs 50a & b). The pulse of sound from the guns radiates out as a hemispherical *wave front*, a portion of which will be reflected back towards the hydrophones from rock interfaces (fig 50a). The path of the minute portion of the reflected wave-front intercepted by a hydrophone group is called a *ray path*. Hydrophone groups spaced along the streamer pick out ray paths that can be related to specific points on the reflector

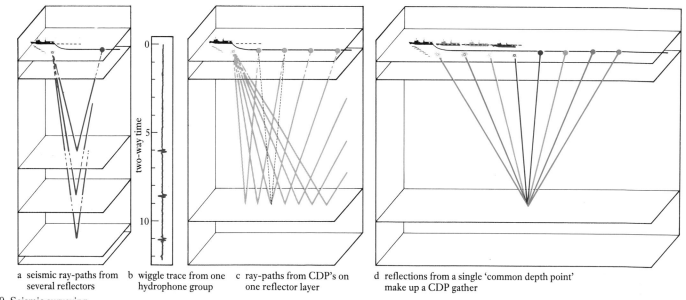

a seismic ray-paths from several reflectors

b wiggle trace from one hydrophone group

c ray-paths from CDP's on one reflector layer

d reflections from a single 'common depth point' make up a CDP gather

50 Seismic surveying

24

surface (fig 50c). Graphs of the intensity of the recorded sound plotted against the two-way time are displayed as *wiggle traces* (fig 50b).

Seismic recording at sea always uses the *common depth point* (CDP) method (figs 50c & d). A sequence of regularly spaced seismic shots is made as the survey vessel accurately navigates its course. Shots are usually timed to occur at distances equal to the separation of the hydrophone groups. In this way up to 120 recordings of the echoes from any one of 240 reflecting points can be collected. Each represents sound which has followed a slightly different ray path, but has all been reflected from the same common depth point.

Processing recordings involves many stages of signal processing and computer summing. First, wiggle traces from a single CDP are collected into groups. Displayed side by side in sequence they form a *CDP gather* (figs 51a & b). Reflections from any one reflector form a hyperbolic curve on the gather because the sound takes longer to travel to the more distant hydrophones. This effect is called *normal move out (NMO)*. Correction is needed to bring the pulses to a horizontal alignment, as if they all came from vertically below the sound source (fig 51c). The separate wiggle traces are added together, or *stacked* (fig 51d). Stacking causes true reflection pulses to enhance one another, and hopefully, random noise will cancel out. This process is repeated for all the CDPs on the survey line. The stacked and corrected wiggle traces are displayed side by side to give a *seismic section* (fig 51e). Most seismic sections used by the oil industry are time-sections (fig 55a) that have undergone a long sequence of data-processing steps designed to improve the quality of the reflections and bring out subtle geological features. For particular purposes, after the principal reflectors have been identified or 'picked', a time-section may be

52 Marine seismic survey in progress

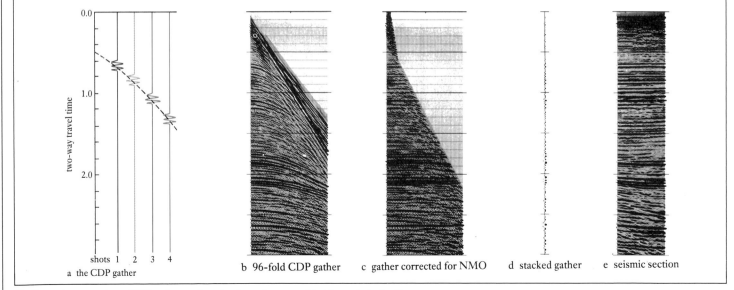

a the CDP gather
shots 1 2 3 4

b 96-fold CDP gather   c gather corrected for NMO   d stacked gather   e seismic section

51 Stages in processing a CDP gather, and a seismic section assembled from stacked gathers

DISCOVERING THE UNDERGROUND STRUCTURE

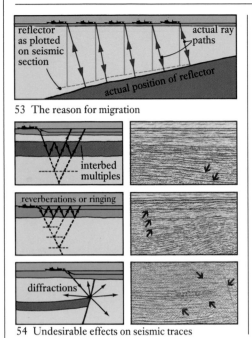

53 The reason for migration

54 Undesirable effects on seismic traces

converted to a depth-section (fig 55b). For this and also for NMO corrections before stacking (p 25), the velocities of sound in the rock layers traversed by the section need to be known. Computer analysis of traces during NMO corrections yields velocity values, but more accurate data comes from special velocity surveys carried out in wells in conjunction with sonic logging (p 31).

Data processing lessens the impact of various undesirable effects that obscure the reflected signals; it also compensates for some intrinsic deficiencies of the CDP method. Undesirable effects (fig 54) include *multiples*, where the sound is reflected repeatedly within a rock formation and, because this takes time, registers as a deeper reflector; reflections between the water surface and the sea-bed are a similar phenomenon known as *ringing*. *Diffractions* are hyperbolic reflections from the broken end of a reflector; they mimic arched formations. Random *noise*, mainly unwanted

reflections from within rock layers, horizontally propagated and refracted sound, bubble pulsations from the airguns and other effects also need to be reduced. Stacking reduces multiples and random noise, but the main computer processing steps are deconvolution, migration, muting and filtering. *Deconvolution* ('decon') aims to counteract the blurring of reflected sound by 'recompressing' the sound to the clean 'spike' emitted from the source. The result is clearer reflections and the suppression of multiples. *Migration* corrects distortions caused by plotting inclined reflectors as if they were horizontal and vertically below the midpoint between shot and receiver; it also collapses diffractions (fig 53). *Muting* cuts out parts of traces embodying major defects such as non-reflected signals; *filtering* removes undesirable noise to enhance the best reflections.

Seismic sections provide 2-dimensional views of underground structure. By using

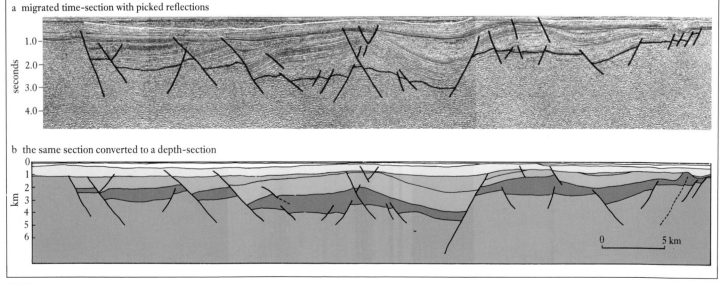

a migrated time-section with picked reflections

b the same section converted to a depth-section

55 Seismic sections

special shooting techniques such as spaced airgun arrays or towing the streamer slantwise, or by shooting very closely spaced lines, it is possible to produce 3-dimensional seismic images. These images (fig 59) comprise vertical sections and horizontal sections ('time-slices'). They are often used in interactive interpretation using a computer and videoscreen. 3-D seismic is expensive to acquire and process; it is normally used to provide precise detail in oilfield development.

Seismic stratigraphy is the study of the depositional interrelationships of sedimentary rock as deduced from an interpretation of seismic data; it can be used in finding subtle sedimentary traps involving changes in porosity. 'Bright-spots', short lengths of a reflection that are conspicuously stronger than adjacent portions may indicate gas: the velocity of sound is sharply reduced in gas-bearing rock, producing a strongly reflective contrast. A gas-water or gas-oil interface may stand out as a noticeably flat reflection amongst arched reflections (fig 56).

The end-products of seismic surveys are interpreted sections showing geological structure down to fine sedimentary details, maps of strongly reflective surfaces identified with known rock units and 'isopach' maps showing the thickness of these units. For the maps, reflections are 'picked' and their depths at points along parallel and intersecting survey lines plotted and contoured. Some rock layers produce wiggles with a distinctive character than can be followed right across a section; others may be identified by comparison with synthetic 'seismograms' made from logging and velocity surveys in existing wells in which the rock sequence is known. The seismic maps are used to identify structures that would either repay more detailed seismic surveying or would warrant wildcat drilling. More detailed surveying would improve the definition of sedimentary and tectonic structures before appraisal wells are drilled.

59 3-D seismic image

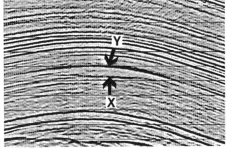

56 'Flat spot' (X) and 'bright spot' (Y)

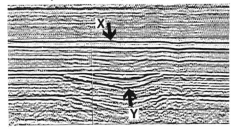

57 Distinctive rock layer (X) and subtle trap (Y)

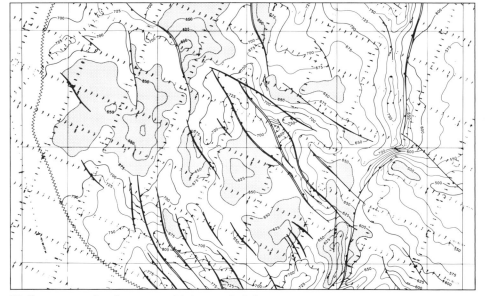

58 Contour map of a reflector; potential traps are shaded

DRILLING

60 Jack-up drilling rig

61 Semi-submersible drilling rig

Drilling rigs are basically of two types: *jack-up* rigs used in shallow waters less than 100 metres deep (fig 60) and *semi-submersible* rigs used in deeper waters down to 360 metres or more (fig 61). In very deep waters, drilling ships are used. Jack-up rigs have lattice legs which are lowered to the sea-bed before the floating section carrying the derrick is raised above the sea surface. Semi-submersible rigs float at all times, but when in position for drilling are anchored and ballasted to float lower in the water with their pontoons below wave-level. Some have 'dynamic-positioning' propellers and can drill in very deep water.

The drilling rig itself is a *derrick* towering above the *drill floor* where most of the human activity is concentrated. The derrick supports the weight of the *drillstring* which is screwed together from 9-metre lengths of *drillpipe*. Hoisting equipment in the derrick can raise or lower the drillstring up to three pipe lengths. At the bottom of the drillstring is a *drill bit* (fig 64), which can vary in size and type. It is attached to the *drill collars*, heavy pipe-sections that put weight on the bit. The rest of the drillstring is supported in tension by the derrick; otherwise it would collapse under its own weight. On semi-submersible rigs, a *compensator* keeps the drillstring stationary while the rig and derrick move. The drill bit is rotated either by turning the whole drillstring or by a downhole turbine driven by the drilling fluid. The latter is a thin mud with a water or oil base which is pumped at high pressure down the hollow drillstring. It lubricates the bit, washes up rock cuttings and most importantly, balances the pressure of fluids in the rock formations below to prevent blowouts or 'gushers'.

In offshore drilling, the first step is to put down a wide-diameter *conductor* pipe to the sea-bed to guide the drilling and contain the drilling fluid. It is drilled into the sea-bed from semi-submersible rigs but on production platforms a pile-driver may be used.

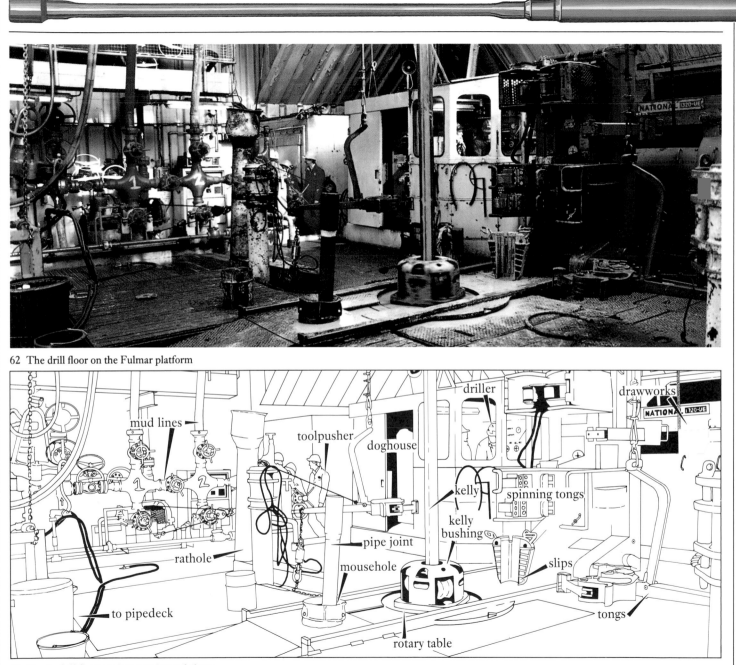

62 The drill floor on the Fulmar platform

63 Key to drill floor equipment pictured above

Labels in figure 63: mud lines, driller, drawworks, NATIONAL 1320-UE, toolpusher, doghouse, kelly, spinning tongs, kelly bushing, pipe joint, rathole, mousehole, slips, to pipedeck, tongs, rotary table

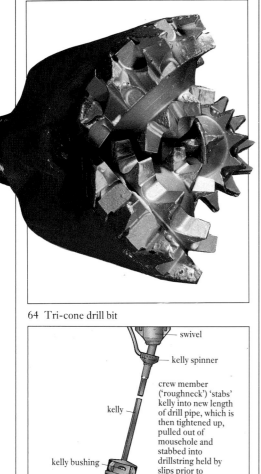

64 Tri-cone drill bit

The two main tasks in drilling are: (1) adding fresh lengths *(joints)* of drillpipe as the drill bit bites down into the rock; and (2) the extraction of the entire drillstring to change the bit or retrieve rock cores (p 31). Withdrawing the drillstring is known as *tripping out* and the whole operation of extraction and re-insertion *(tripping in)* of the drillstring is a *round trip*. In rotary drilling, the rotary motion is imparted to the drillstring by the power-driven *rotary table* in the drill floor. The rotary table is coupled to a square-sectioned pipe known as the *kelly* via the *kelly bushing*. The kelly is screwed to the drillstring, and the bushing allows it to drill down. To add a pipe joint the drillstring is raised, clamped in the rotary table with wedges *(slips)* and the kelly unscrewed *(broken out)*. The kelly, which is suspended from the *swivel*, is moved over (fig 65) to a fresh pipe joint which has meanwhile been hauled up from the pipe deck and placed ready in the *mousehole*. It is *stabbed* into the new pipe joint and *spun up* with the *kelly spinner*. The new joint is lifted out of the mousehole, stabbed into the drillstring, spun up and then tightened up with the *tongs* or nowadays with the *spinning* or *power tongs* (fig 63). The drillstring is then lowered and drilling restarted. The raising and lowering of the travelling block and the maintenance of correct tension on the drillstring is controlled by the *driller* operating the *drawworks lever* in the *doghouse*, which also houses much other rig instrumentation.

For tripping out, the drillstring is clamped with the slips; the kelly is broken out and the kelly, kelly bushing and swivel placed in the *rathole* (fig 66). The travelling block is lowered and the *elevators* (attached to the hook on the travelling block) clamped under the drillpipe *joint box*. A stand of pipe made up of three joints is withdrawn, the drillstring clamped and the stand unscrewed with the spinning tongs. The stand is then removed from the elevators and stacked vertically in the derrick by the *derrickman* standing on the *monkeyboard* high up in the derrick. Finally the bit is removed, using a *bit breaker* placed in the rotary table. *Tripping in* is simply the reverse of tripping out. The round trip may take several hours to accomplish.

Completed sections of wells are cased with steel pipe cemented into place. To the top of the casing is attached the *blowout preventer*, a stack of hydraulic rams which can close off the well instantly if backpressure (a *kick*) develops from invading oil, gas or water. A more common problem is the drillstring sticking in difficult rock formations such as the thick Tertiary clays in the North Sea. A hydraulic device known as a *jar*, mounted between the drill collars, can give the drillstring a series of jolts. If that doesn't work, other techniques may be used, including spotting with oil or water. Special *fishing* tools can also retrieve stuck pipe and broken equipment *(junk)*.

swivel

kelly spinner

crew member ('roughneck') 'stabs' kelly into new length of drill pipe, which is then tightened up, pulled out of mousehole and stabbed into drillstring held by slips prior to resumption of drilling.

kelly

kelly bushing

pipe joint

mousehole

drillstring clamped by slips

slips

drilling floor

rotary table

65 Adding a new joint of pipe

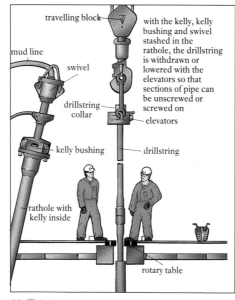

travelling block

mud line

swivel

drillstring collar

kelly bushing

with the kelly, kelly bushing and swivel stashed in the rathole, the drillstring is withdrawn or lowered with the elevators so that sections of pipe can be unscrewed or screwed on

elevators

drillstring

rathole with kelly inside

rotary table

66 Tripping

GETTING THE MOST OUT OF A WELL

Drilling with tri-cone bits (fig 64) grinds up the rock into tea-leaf-sized cuttings which are brought to the surface by the drilling mud. The drilling mud is passed over a *shale shaker* which sieves out the cuttings. In exploration drilling, the cuttings are taken for examination by a geologist known as a *mudlogger* who is constantly on the lookout for oil and gas. Oil entrapped in the mud is detected by its fluorescence in UV light. Gas is extracted from the mud in a gas trap and sent under vacuum to a gas detector and analyser. An increase in the amount triggers an alarm to alert the mudlogger and the drilling superintendent. If laboratory tests are needed on potential reservoir rock, a solid *core* of rock can be drilled by a special hollow drilling bit. Each short length of core retrieved calls for a 'round trip', so coring is an expensive operation not undertaken lightly.

After a well is drilled but before it is cased, vital information is gained by lowering measuring devices down the hole on a wireline. This is generally carried out by specialist contractors while the rig is being set up for casing and cementing (fig 68). The measuring devices are contained in a *sonde*; the data is transmitted electronically to the surface via the wireline as the sonde is steadily pulled up the hole. The result is a *wireline log* (fig 67). Various types of sonde measure: (1) electrical resistivity including spontaneous potential; (2) the speed of sound in adjacent rock; (3) backscatter of gamma rays from a gamma source in the sonde; (4) gamma rays from hydrogenous materials bombarded by neutrons from a source in the sonde; and (5) natural gamma radiation of the rocks. The data from *sonic* (2), gamma backscatter or *formation density* (3) and *neutron* (4) logs give indications of porosity and hence permeability. Porosity in combination with *resistivity* (1) give indications of hydrocarbons.

Other devices measure hole diameter, dip of strata and the direction of the hole. Sidewall corers which punch or drill out small cores of rock, and geophones for well velocity surveys and seismic profiling are also lowered into uncased wells. In deviated wells approaching the horizontal, flexible high-pressure steel *coiled tubing* may be used to carry wireline logging tools and for performing wellbore maintenance operations.

If hydrocarbons have been detected in a well, a *repeat formation tester* is lowered on a wireline. This measures fluid pressures and collects small samples. Full-scale *production tests* in the North Sea are only carried out on cased wells. Production tubing with valves and a packer is lowered into the hole to seal off the interval to be tested and the casing below the packer is perforated. The well is allowed to flow for several hours while pressure and flow rates are measured. These results are taken into account in the siting of any follow-up appraisal wells.

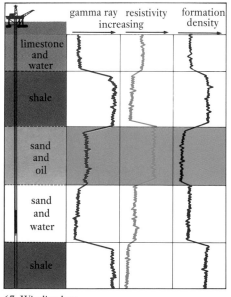

67 Wireline logs

68 Repeat formation tester

DEVELOPING A DISCOVERY

When promising amounts of oil or gas are found in a wildcat well, a programme of detailed field appraisal may begin. The size of the field must be established, and the most efficient production method worked out in order to assess whether it will repay, with profit, the huge costs of offshore development and day-to-day operation. Appraisal may take several years to complete and is itself very costly.

Appraisal draws together information from all available techniques. Detailed seismic surveys build up an accurate 3-dimensional image of the discovery (p 27), and appraisal wells are drilled to confirm the size and structure of the field (fig 69). Wireline logging in each new well yields data on porosity and fluid saturation and the thickness of the hydrocarbon-bearing rocks, while production testing yields hydrocarbon samples and information on reservoir productivity, temperatures and pressures (p 31). Oil, gas and reservoir rock samples are analysed in the laboratory. Most fields have both good and bad features which must be fully considered when deciding whether to develop (pp 36–37).

Production may prove difficult and expensive if the reservoir rock is seriously disrupted by faulting or contains extensive areas of poor permeability. Porosity and permeability may vary dramatically where the reservoir rock consists of a variety of sediments (figs 72–74), and may be much reduced in areas where later mineral growth blocks the available pore spaces (p 4, fig 9). Geologists compare core samples from the deeply buried reservoir rock with present-day sediments to identify the environment in which it accumulated. This environment is used to develop a geological model to help predict likely variations in the reservoir rock types and properties. If, for example, the best-quality reservoir rock is a dune sand or a beach sand, its likely extent and thickness can be estimated from the size and shape of a comparable modern dune complex or beach.

The identification of microfossils that inhabited particular environments, such as shallow seas or brackish lagoons, helps confirm the model, as well as indicating the age of the reservoir rock (figs 70 & 71). Geologists and reservoir engineers use the geological model to select the best sites for production wells.

Studies in the Brent Field showed that the reservoir rocks most closely resemble the sediments deposited in a large delta. This geological model explained the interlayering of muddy, poorly permeable rocks with better-quality reservoir sandstones. Fig 73 shows the varied delta environments where these rocks may have accumulated, and fig 74 indicates the environments that produced the better-quality reservoir rocks. It also indicated the possibility of finding more oilfields within this ancient delta beneath the northern North Sea. Further exploration proved this to be so; these deltaic rocks are the most prolific oil reservoirs in the North Sea.

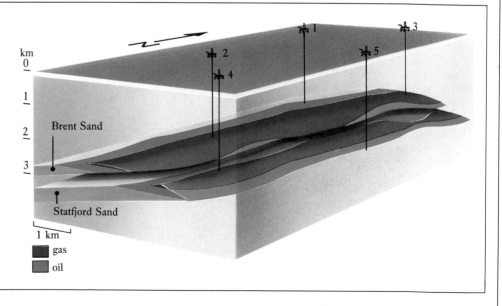

1 discovery well

2 proved that the field extends southwards

3 proved that the field extends northwards

4 discovered a second, deeper reservoir

5 dry well indicated eastern limit of field

km
0
1
2
3

Brent Sand

Statfjord Sand

1 km

gas

oil

69 Delineating the Brent Field reservoirs

70 Dinoflagellate (microplankton) ×400

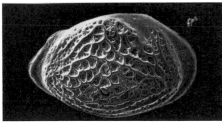

71 Near-shore marine ostracod (crustacean) ×70

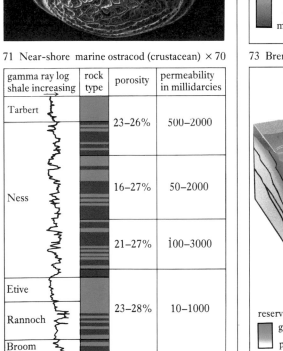

gamma ray log shale increasing	rock type	porosity	permeability in millidarcies
Tarbert		23–26%	500–2000
Ness		16–27%	50–2000
		21–27%	100–3000
Etive		23–28%	10–1000
Rannoch			
Broom			

72 Reservoir properties in the Brent Field

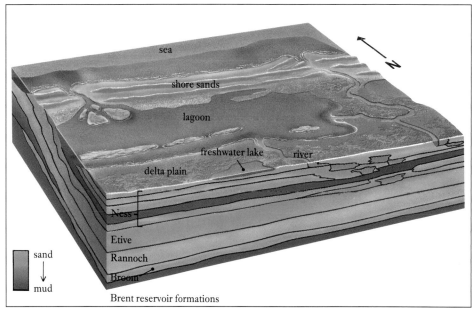

sand
↓
mud

Brent reservoir formations

73 Brent Delta environments and rock types

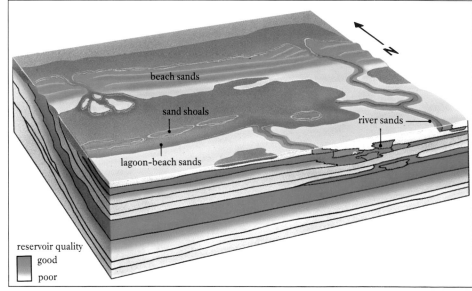

reservoir quality
good
poor

74 Reservoir quality of the Brent Delta rocks

HOW MUCH OIL AND GAS?

When deciding whether to develop a field, a company must estimate how much and how easily oil and gas will be recovered. Although the volume of oil and gas in place can be estimated from the volume of the reservoir, its porosity, and the amount of oil or gas in the pore spaces (fig 77), only a proportion of this amount can be recovered. This proportion is the *recovery factor*, and is determined by various factors such as reservoir dimensions, pressure, the nature of the hydrocarbon, and the development plan.

Pressure is the driving force in oil and gas production. *Reservoir drive* is powered by the difference in pressures within the reservoir and the well (fig 76), which can be thought of as a column of low surface pressure let into the highly pressured reservoir. If permeability is good and the reservoir fluids flow easily, oil, gas and water will be driven by natural depletion into the well and up to the surface. Expansion of the gas cap and the water drives oil towards the well bore. Gas and water occupy the space vacated by the oil. In reservoirs with insufficient natural drive energy, water or gas is injected to maintain the reservoir pressure (p 43).

The proportion of oil that can be recovered from a reservoir is dependent on the ease with which oil in the pore spaces can be replaced by other fluids like water or gas. Tests on reservoir rock in the laboratory indicate the fraction of the original oil in place that can be recovered. Viscous oil is difficult to displace by less viscous fluids such as water or gas as the displacing fluids tend to channel their way towards the wells, leaving a lot of oil in the reservoir. The quoted recovery factor for most North Sea fields is about 35 percent, but may be as low as 9 percent where the oil is very viscous, or perhaps as high as 70 percent where reservoir properties are exceptionally good and the oil of low viscosity. The recovery factor in gasfields is much higher, figures of over 85 percent being quoted for most.

Each oil and gas reservoir is a unique system of rocks and fluids that must be understood before production is planned. Petroleum engineers use all the available data to develop a mathematical model of the reservoir. Computer simulations of different production techniques are tried on this *reservoir engineering model* to predict reservoir behaviour during production, and select the most effective method of recovery. For example, if too few production wells are drilled water may 'cusp' or channel towards the wells, leaving large areas of the reservoir unswept.

Factors such as construction requirements, cost inflation, future oil prices and possible changes in Government tax laws must also be considered when deciding whether to develop an oil- or gasfield (pages 36–37). When a company is satisfied with the plans for development and production, they must be approved by the Department of Energy, which monitors all aspects of offshore development.

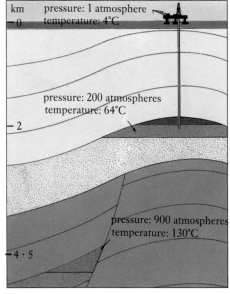

75 Reservoir conditions vary with depth

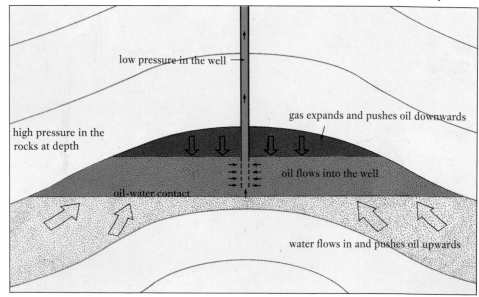

76 Reservoir drive

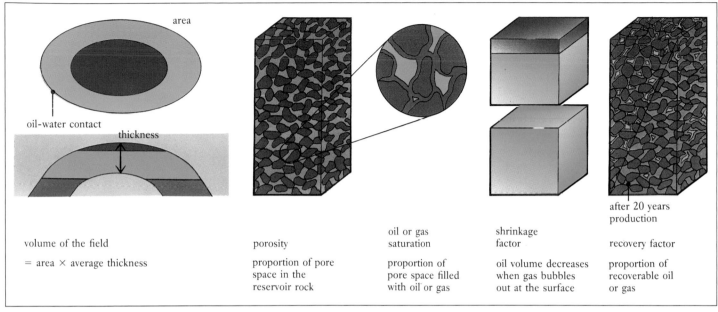

volume of the field

= area × average thickness

porosity

proportion of pore space in the reservoir rock

oil or gas saturation

proportion of pore space filled with oil or gas

shrinkage factor

oil volume decreases when gas bubbles out at the surface

after 20 years production

recovery factor

proportion of recoverable oil or gas

77 Factors determining the recoverable quantity of hydrocarbons from a field

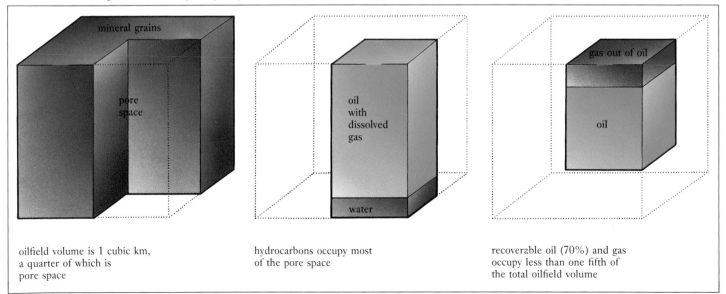

oilfield volume is 1 cubic km, a quarter of which is pore space

hydrocarbons occupy most of the pore space

recoverable oil (70%) and gas occupy less than one fifth of the total oilfield volume

78 Possible recoverable oil from the most productive North Sea oilfield

PATHWAYS TO PRODUCTION

Discovering and developing a profitable oilfield is a long, complex and costly business. Many factors may speed or delay progress, from the local North Sea weather to the state of the international oil market, as you will discover!

To play, you need one dice and a counter for each player. Throw the dice in turn and move forward by the number of steps shown, EXCEPT where told to do otherwise.

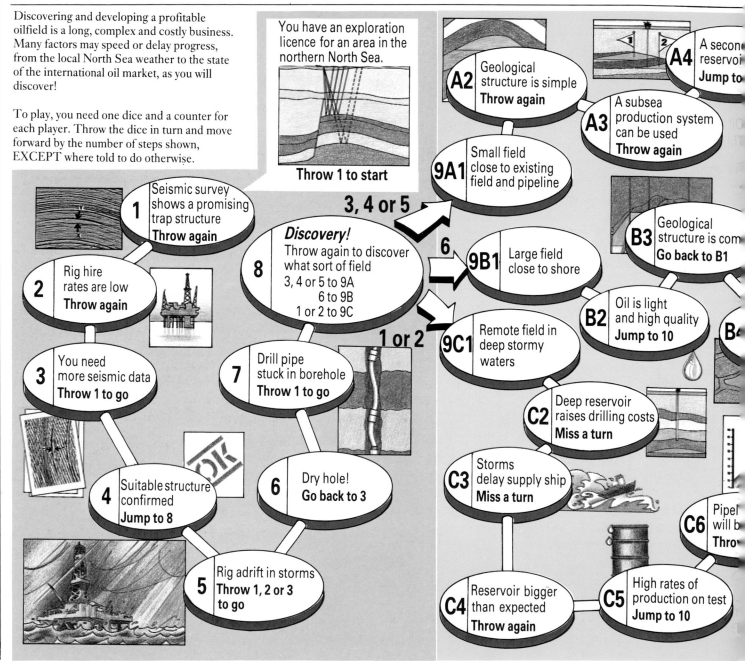

You have an exploration licence for an area in the northern North Sea.

Throw 1 to start

1 Seismic survey shows a promising trap structure **Throw again**

2 Rig hire rates are low **Throw again**

3 You need more seismic data **Throw 1 to go**

4 Suitable structure confirmed **Jump to 8**

5 Rig adrift in storms **Throw 1, 2 or 3 to go**

6 Dry hole! **Go back to 3**

7 Drill pipe stuck in borehole **Throw 1 to go**

8 *Discovery!* Throw again to discover what sort of field
3, 4 or 5 to 9A
6 to 9B
1 or 2 to 9C

3, 4 or 5

6

1 or 2

9A1 Small field close to existing field and pipeline

9B1 Large field close to shore

9C1 Remote field in deep stormy waters

A2 Geological structure is simple **Throw again**

A3 A subsea production system can be used **Throw again**

A4 A second reservoi... **Jump to**

B2 Oil is light and high quality **Jump to 10**

B3 Geological structure is com... **Go back to B1**

B4

C2 Deep reservoir raises drilling costs **Miss a turn**

C3 Storms delay supply ship **Miss a turn**

C4 Reservoir bigger than expected **Throw again**

C5 High rates of production on test **Jump to 10**

C6 Pipel... will b... **Thro...**

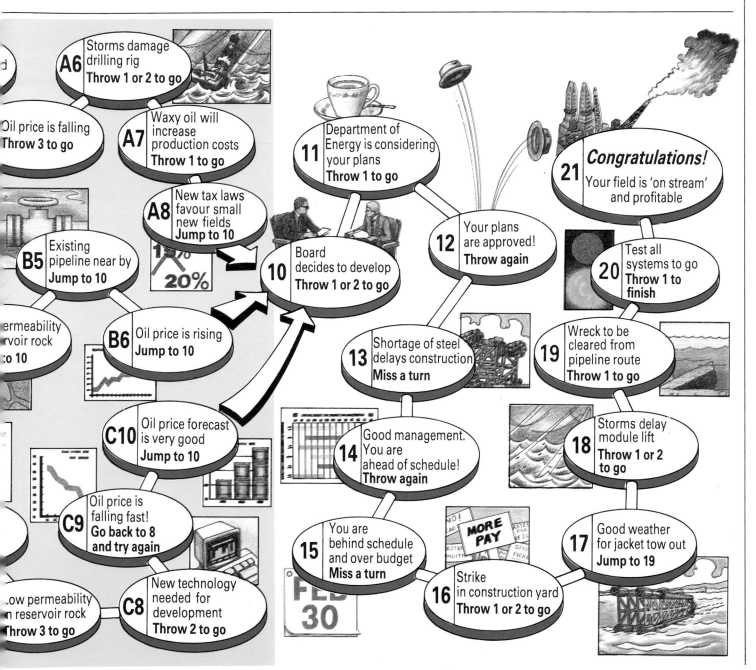

THE OFFSHORE CHALLENGE

When development of the North Sea fields began in the mid-60s, the industry had never before faced such a hostile environment. Whilst simple platform designs derived from those used in the Gulf of Mexico sufficed for the shallow southern North Sea, the severe storms and great water depths of the northern North Sea called for major engineering and technological innovation. Production facilities had to be designed to withstand wind gusts of 180 km/hour and waves 30 metres high (fig 81). Other problems included the ever-present salt-water corrosion (fig 79) and fouling by marine organisms. Dealing with the many underwater construction and maintenance tasks falls to divers without whom the industry would grind to a halt. Giant floating cranes designed to lift ever greater loads were commissioned and many other specialized craft had to be developed to establish and service the offshore industry. Huge helicopter fleets were needed to ferry workers to and from the platforms and rigs.

79 Salt-water corrosion on a North Sea platform

80 Diver inspecting subsea platform structure

81 Brent Charlie platform in rough seas

Most oil and gas production platforms in offshore Britain rest on steel supports known as 'jackets', a term derived from the Gulf of Mexico. A small number of platforms are fabricated from concrete. The steel jacket, fabricated from welded pipe, is pinned to the sea floor with steel piles. Above it are prefabricated units or modules providing accommodation and housing various facilities including gas turbine generating sets. Towering above the modules are the drilling rig derrick (two on some platforms, p 1, fig 1), the flare stack in some designs (also frequently cantilevered outwards) and service cranes. Horizontal surfaces are taken up by store areas, drilling pipe deck and the vital helicopter pad. The eight-legged steel jacket for the North West Hutton field stands 145 metres tall, weighs 15 000 tonnes and supports 20 000 tonnes of equipment. The entire structure is 230 metres from the sea-bed to the top of the derricks.

Concrete gravity platforms are so-called because their great weight holds them firmly on the sea-bed. They were first developed to provide storage capacity in oilfields where tankers were used to transport oil, and to eliminate the need for piling in hard sea-beds. The Brent D platform, which weighs more than 200 000 tonnes, was designed to store over a million barrels of oil. But steel platforms, in which there have been design advances, are now favoured over concrete ones.

Several platforms may have to be installed to exploit the larger fields, but where the capacity of an existing platform permits, subsea collecting systems linked to it by pipelines have been developed using the most modern technology. They will be increasingly used as smaller fields are developed. For very deep waters, one solution is the Hutton Tension Leg Platform: the buoyant platform, resembling a huge drilling rig, is tethered to the sea-bed by jointed legs kept in tension by computer-controlled ballast adjustments.

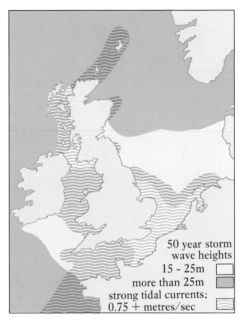

50 year storm wave heights
15 - 25m □
more than 25m ▨
strong tidal currents;
0.75 + metres/sec ▨

83 Some offshore hazards

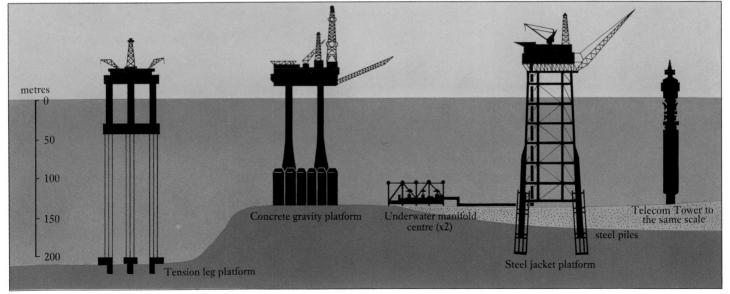

metres
0
50
100
150
200

Concrete gravity platform

Underwater manifold centre (x2)

Tension leg platform

Steel jacket platform

steel piles

Telecom Tower to the same scale

82 Production installations in the central and northern North Sea

CONSTRUCTION & INSTALLATION

The scale of offshore oil and gas construction projects is vast, especially for the oilfields of the northern North Sea. The large fields discovered in the early 1970s took an average of five years from the beginning of development to the date of production start-up, and each cost over a billion pounds in 1987 prices. Between 1969 and 1987, the industry spent about 47 billion pounds (in 1987 prices) on offshore development. Since 1980 three-quarters of this expenditure has been made in Britain.

As soon as field appraisal has shown that development would be a commercial success, orders for one or more platforms and associated pipelines go out. There are several yards in Scotland which specialize in jacket fabrication, which can take up to two years to complete. The completed jacket must be towed out to the field in calm weather, usually during the 'fine weather window' of the summer months (fig 84). It is launched off its barge and up-ended into position by the controlled flooding of ballast tanks in its legs. After piles are driven to secure the jacket to the sea-bed, barges bring the deck support and production modules to be lifted into position on the jacket (fig 86). Specially designed crane vessels can lift over 10 000 tonnes. At this stage of the project, rough weather can cause serious delays. Concrete platforms are built in deep, sheltered fjords or sea lochs. As new concrete is poured, the structure gradually sinks (fig 85). The deck and modules are placed on the legs close to shore, then the platform is towed out to the field. With ballast water pumped into the storage tanks, the platform settles firmly on the sea-bed. Inside the platform, the hook-up and testing of equipment ready for drilling and production may take another season, and require up to a thousand installation workers, called 'bears', at any one time. Finally, after completion of the first of the wells, the platform comes on-stream, beginning a producing life of at least twenty years.

85 Brent D concrete platform under construction

84 Towing out the steel jacket of the North West Hutton platform

86 Hoisting a platform module, NW Hutton Field

FUNCTIONS OF A PRODUCTION PLATFORM

Oil platforms are industrial towns at sea, carrying the personnel and equipment needed for continuous hydrocarbon production. The most important functions are drilling, preparing water or gas for injection into the reservoir, processing the oil and gas before sending it ashore, and cleaning the produced water for disposal into the sea. Power is generated on the platform to drive production equipment and support life. All production systems are constantly monitored for leaks, since oil and gas are hazardous and extremely flammable.

The top end of each production well sprouts a branching series of pipes, gauges and valves called the 'Christmas tree' (p 42, fig 89). At this point, crude oil is a hot, frothy, corrosive, high-pressure fluid containing gas, water and sand. After separation, the crude oil is metered and pumped into the pipeline, or stored until sent ashore by tanker. The gas separated from the oil may be used for fuel, or compressed and piped to shore or re-injected into the reservoir. Any gas that cannot be used or piped ashore must be burnt in the platform's flare. Very little gas is now flared. Processing systems for the gasfields of the southern North Sea are relatively simple. The gas liquids are removed, then the gas is compressed, cooled, dehydrated and metered before being piped to shore.

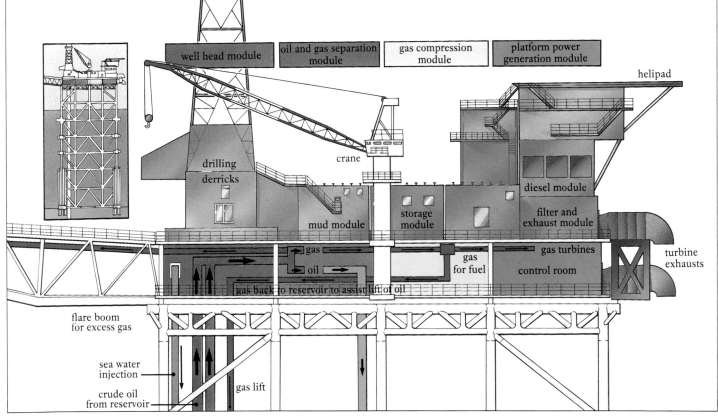

87 Functions of a production platform

PRODUCTION WELLS

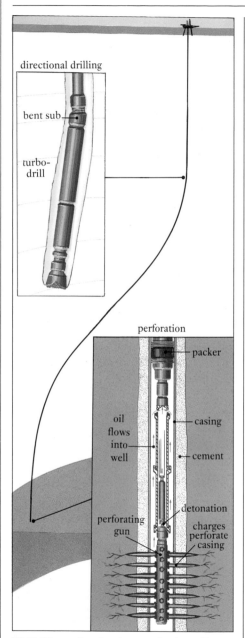

directional drilling

bent sub

turbo-drill

perforation

packer

oil flows into well

casing

cement

perforating gun

detonation

charges perforate casing

To develop offshore fields as economically as possible, numerous directed wells radiate out from a single platform to drain a large area of reservoir (fig 91). Forties, the largest UK oilfield, has over 120 wells from five platforms. For directional drilling, weighted drill collars deflect the bit by a specific angle. Turbo-drilling (p 28) with a 'bent sub' is another technique where only the deflected bit rotates. Thereafter the driller uses various techniques to keep the hole on course. Directionally drilled wells have achieved angles of more than 60 degrees from vertical. The slant-drilling technique, where wells are drilled at an angle from the surface, is used in the Morecambe Bay gasfield to tap a shallow and extensive reservoir. When drilling through the reservoir rock, it is important that the mud filtrate does not damage the reservoir permeability. To overcome this, drillers have used oil-based mud when drilling through many North Sea reservoirs. Oil-based mud also speeds drilling through

shales, an important factor in building up production rates. Crude hydrocarbons contain acidic fluids like CO_2 which could corrode casing, so steel production tubing is inserted into the well to collect oil and gas and protect the casing.

As the well is prepared for completion, a perforating gun with shaped charges is set below the bottom of the tubing. Packers above the gun will prevent leakage of oil or gas into the space between the tubing and casing. At a signal from the platform above, the charges perforate the casing, cement and reservoir rock, allowing oil and gas to enter the wellbore. From now on, flow will be controlled from the valves on the Christmas tree at the wellhead (fig 89). After the drilling and completion of each well, the derrick can be moved on skids to the next well slot on the platform. Eventually, rows of Christmas trees mark the location of all the production wells. Increasingly, completions in smaller fields are located on the sea-bed.

88 Drilling and perforating a production well

89 'Christmas trees' on an offshore production platform

To achieve as high a recovery factor as possible in offshore oilfields, reservoir pressures must not be allowed to fall too low as oil and associated gas are removed. It is desirable to maintain pressures above the point where dissolved gas in the oil comes out of solution to form free gas. Seawater is pumped into the water-soaked rocks beneath the oil zone in volumes equal to the sub-surface volume of the liquids produced. Water injection wells are usually located around the periphery of an oilfield. Gas separated from oil on the platform may also be compressed and injected into the reservoir rocks to maintain pressure. Water and gas injection can improve recovery of oil from less than 15 percent to more than 50 percent. Very deep fields, such as Brae, with high pressures and temperatures may yield condensate, a valuable light oil which exists as gas vapour in the reservoir. Dry gas will be injected into the reservoir to maintain pressure, avoiding condensate 'drop out' and sweeping the rich gas to the wells. Downhole pumps have been used offshore when reservoir pressures are insufficient to send the oil to the surface, as in the Beatrice Field. A more common technique is gas lift in which gas from the same or a nearby field is mixed with the oil in the tubing to lessen the weight of the liquid column.

Flow from every oil and gas well is tested and monitored throughout the well's life. Replacement of worn equipment such as tubing and valves helps prolong the life of the well. In less productive wells, well stimulation may be tried. High-pressure fluids are pumped down the well to create deep fractures in the reservoir rock through which oil or gas can flow. These fractures are held open by sand grains which are forced into the fracture with the fluid. Acid stimulation helps remove clogging mineral scale such as calcium carbonate which may have accumulated during years of production.

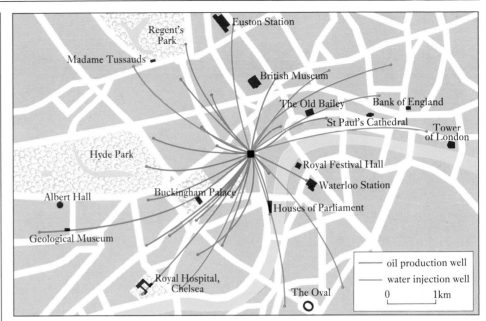

90 The well system below superimposed on Central London

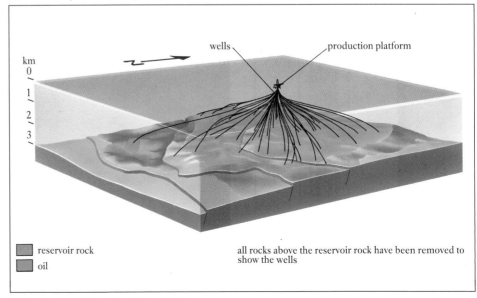

91 Deviated wells in a North Sea oilfield

43

WORKING OFFSHORE

92 Emergency Support Vessel *Iolair*

93 Lifeboat drill on Ninian Central platform

In 1984 and 1985 during the peak of oil production, about 30 000 workers were employed in UK offshore fields. For logistical support, the offshore operating companies have established onshore supply bases which communicate with the platforms, transfer personnel and ensure delivery of food and equipment. Production continues with the minimum of delays, despite the hostile North Sea weather. The Forties Field, with one unmanned and four manned platforms, has a staff of 643 offshore and 56 onshore. Among other consumables, supply boats bring to Forties each year 100 000 litres of fresh milk, 50 tonnes of beef, 130 tonnes of potatoes and 12 000 litres of cooking oil. All rubbish is brought back to shore for disposal. Living quarters are compact but comfortable. Food is good and abundant and the ship is 'dry'. Off-shift, a worker can choose to work out in the gym, watch a video, play snooker, browse through the library or learn to use a personal computer. With fourteen-day stints of twelve-hour shifts on a remote platform, an offshore worker requires stamina and the ability to cooperate in a group. The Offshore Installation Manager of a platform is like the captain of a ship, making sure that all operations run smoothly and safety standards are met. He coordinates the work of different groups such as drilling, production and maintenance, and communicates progress or problems to 'the beach'.

Safety is always the foremost consideration. Personnel are provided with protective clothing and must participate in regular safety drills. In the event of a serious emergency, specialized vessels are always on call. *Iolair*, BP's Emergency Support Vessel, can direct 50 000 gallons of water a minute to control a platform fire, and is equipped to deal with blow-outs. It also has hospital facilities, including an operating theatre. For routine platform inspection and maintenance, *Iolair* provides diving support and a workshop.

Most offshore oil and all offshore gas is brought to shore by pipelines which operate in all weathers. Pipeline routes are planned to be as short as possible. Slopes that could put stress on unsupported pipe are avoided and sea-bed sediments are mapped to identify unstable areas and to see if it will be possible to bury the pipe. Pipeline construction begins onshore, as lengths of pipe are waterproofed with bitumen and coated with steel-reinforced concrete. This coating weighs down the submarine pipeline even when it is filled with gas. The prepared pipe-lengths are welded together offshore on a laybarge. As the barge winches forward on its anchor lines, the pipeline drops gently to the sea-bed, guided by a 'stinger'. Pipelines need to be cleaned regularly to remove wax deposits and water: a collecting device known as a pig is forced through the pipe.

Where tankers transport oil from small or isolated fields, various oil storage systems are used, such as cylindrical cells in the concrete Beryl platform and in the steel Maureen platform, or the conversion of a large crude tanker to a Floating Storage Unit in the Fulmar Field. Tankers cannot be manoeuvred close to a platform, so the oil is piped to a buoy to which a specially adapted tanker is moored. The larger tankers load oil from this Floating Storage Unit. At the Maureen Field, oil is piped to an articulated loading column over 2 km from the platform (fig 96). It takes about 14 hours to take on board a load of 404 000 barrels at the Beryl Field.

In onshore terminals, carefully landscaped to minimise their environmental impact, crude oil and gas undergo further processing. Any remaining water and gas are removed from oil which is stored at the terminal before transport to refineries. Gas is dried and then given its characteristic smell before entering the national grid. During transportation, great care is taken to avoid or deal effectively with spillage.

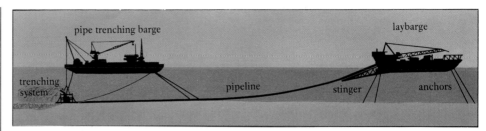

94 Pipeline installation and burial in a soft sea-bed

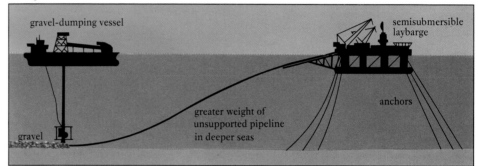

95 Installing a pipeline in deeper seas. On hard sea-beds pipelines are covered

96 Tanker loading oil from an articulated loading column, Maureen Field

ORGANISING OFFSHORE DEVELOPMENT

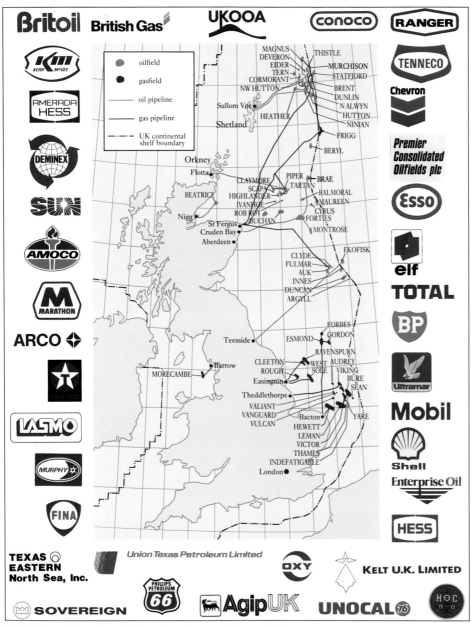

97 Offshore hydrocarbon fields and operators, as at January 1988

In 1964, Britain enacted the Continental Shelf Act to extend sovereignty offshore, drew up the Petroleum Production Regulations to provide a system for licensing and control and issued the first round of Licences. By 1965 Britain had agreed her North Sea boundaries with the Netherlands, Denmark and Norway, and exploration had begun.

For licensing purposes, the United Kingdom continental shelf is divided into quadrants, the areas of one degree latitude by one degree longitude shown in fig 97. Each quadrant is further divided into thirty blocks of approximately 250 square km. Companies are invited to apply for the right to explore blocks selected by the Department of Energy in licensing rounds held every two years. Licences are awarded on a mainly discretionary basis to companies judged to have good operational records, well-prepared exploration and production plans, and financial soundness. The companies must also be registered in Britain.

The offshore petroleum industry generates huge tax revenues and has created many jobs in industries which supply services and equipment. Of the price of a barrel of oil, 54 percent goes to tax and 33 percent pays production costs, leaving 13 percent for the producer. Tax revenues for 1984 and 1985, the peak oil production years, were over 10 billion pounds per year.

The operating companies have formed the United Kingdom Offshore Operators Association (UKOOA) to provide a forum to deal with common interests and problems, and to represent the operators to the Government and other interested groups. These functions are performed through specialist committees which deal with all aspects of offshore operations, including legislative and economic concerns, maintenance of safety standards, provision of co-ordinated offshore emergency services, and pollution prevention and control.

'North Sea Gas' was first discovered in the southern North Sea in 1965 and brought ashore in 1967. The giant Forties Field was discovered in 1970 and the first oil (from the Argyll Field) came ashore in 1975. All the largest and most easily developed oil fields have been discovered and are now past their production peak. Oil production generally peaked in 1985 (fig 100) though gas production was still rising in 1988. Both will last well into the 21st century. Further exploration will continue, searching in particular for 'subtle' traps. If oil prices return to a high level, it may pay to apply 'enhanced recovery' techniques to existing and future finds. These techniques are costly and would require specialized platform facilities. They include injection of steam, surfactant chemicals, miscible gases and polymers to lower the viscosity of the oil and increase flow rates and yields. There will be increasing development of small satellite fields linked by subsea collectors and pipelines to an existing central platform and pipeline to shore.

Ultimately all the extractable oil and gas will be removed and the fields will be abandoned. Much thought and research is going into abandonment procedures and methods that minimize the impact on fishing and the environment, but without incurring excessive costs. Steel platforms may be lopped off to an agreed depth below sea level to permit safe navigation and will become havens of marine life; in deep waters they could be toppled by explosive cutting techniques. Concrete platforms may possibly be refloated and towed away to be sunk in the deep ocean. Pipelines could be flushed with water and left in place on the sea-bed. In all likelihood, each structure will need to be examined individually to determine the optimum solution after the wells have been plugged and the platform facilities cleaned of hazardous products.

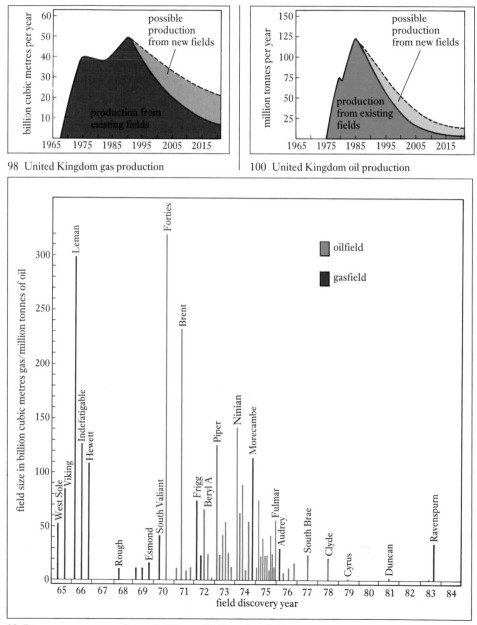

98 United Kingdom gas production

100 United Kingdom oil production

99 Discovery sequence and size of offshore oilfields and gasfields

INDEX

British Library Cataloguing in Publication Data
Geological museum
    Britain's offshore oil and gas.
    1. North Sea. Natural gas deposits and petroleum deposits
    I. British Museum (Natural History)
553.2'8'0916336
ISBN 0-565-01029-8

Authors:
Geological Museum:
Fred Dunning, Ian F Mercer, Pat Taylor, Christine Woodward, Robin Sanderson
Shell UK:
Ken Glennie, Keith Eastwood

Editor: Christine Woodward

Artists: Gary Hincks, Mike Eaton, Bob Chapman, Irwin Technical, M L Design, Panda Art

Designer: David Robinson

Typesetting by Cambridge Photosetting Services
Printed in Great Britain by Jolly & Barber, Rugby

Photographs and diagram material:
Front and back covers, figs 70, 81, 85 Shell U.K. Limited
Inside front cover British Gas
Fig 1 and inside back cover Marathon Oil U.K. Limited
Fig 3 Dr G Boalch
Figs 4, 5 Open University
Figs 8, 9 British Geological Survey
Figs 51, 52, 54, 58, 59 Horizon Exploration Limited
Fig 55 a&b Merlin Geophysical Limited
Figs 56, 57 Seismograph Service (England) Limited
Fig 60 Conoco (U.K.) Limited
Figs 61, 89 Occidental Petroleum (Caledonia) Limited
Fig 62 F W Dunning
Figs 64, 92 The British Petroleum Company
Fig 68 Schlumberger Ltd & Newpoint Publishing Company
Fig 71 University College London
Fig 79 G Velarde
Figs 80, 93 Chevron Petroleum (U.K.) Limited
Figs 84, 86 Amoco (U.K.) Exploration Company
Fig 96 Phillips Petroleum Company U.K. Limited

We should like to thank individuals in many companies who have helped with pictures and information for this booklet. Our especial thanks go to the staff at Shell Exploration and Production, Horizon Exploration Limited, and Merlin Geophysical Limited.